A STUDENT'S GUIDE

TO THE STUDY, PRACTICE,
AND TOOLS
OF MODERN MATHEMATICS

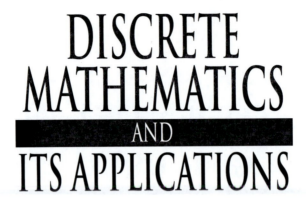

DISCRETE MATHEMATICS AND ITS APPLICATIONS

Series Editor

Kenneth H. Rosen, Ph.D.

Titles *(continued)*

DISCRETE MATHEMATICS AND ITS APPLICATIONS

Series Editor KENNETH H. ROSEN

A STUDENT'S GUIDE

TO THE STUDY, PRACTICE, AND TOOLS OF MODERN MATHEMATICS

Donald Bindner

Truman State University

Kirksville, Missouri, USA

Martin Erickson

Truman State University

Kirksville, Missouri, USA

CRC Press
Taylor & Francis Group
Boca Raton   London   New York

CRC Press is an imprint of the
Taylor & Francis Group, an **informa** business

A CHAPMAN & HALL BOOK

Chapman & Hall/CRC
Taylor & Francis Group
6000 Broken Sound Parkway NW, Suite 300
Boca Raton, FL 33487-2742

© 2011 by Taylor and Francis Group, LLC
Chapman & Hall/CRC is an imprint of Taylor & Francis Group, an Informa business

No claim to original U.S. Government works

Printed in the United States of America on acid-free paper
10 9 8 7 6 5 4 3 2 1

International Standard Book Number: 978-1-4398-4606-3 (Paperback)

Visit the Taylor & Francis Web site at
http://www.taylorandfrancis.com

and the CRC Press Web site at
http://www.crcpress.com

Contents

Preface

When you study mathematics at the university level, there is much to learn. You must learn a new language (the language of mathematics) and several new skills, including the use of mathematical software and other resources. You should also learn about "math culture." Wouldn't it be nice if there was a guidebook that you could turn to for help?

This is that book. Here you can find answers to such questions as: How do I study mathematics? How do I write a mathematical proof? How do I write a mathematical paper? How do I do mathematical research? How do I give a mathematical presentation? These issues, and many others, are discussed in short chapters on each topic. You can read a chapter in its entirety, or just get the facts you need and move on. Either way, you learn something new and clear a hurdle toward doing what you want to do mathematically.

You can get more out of your mathematical studies by knowing how to use mathematical tools. How do I compose a LATEX file? How do I give a math talk using Beamer? How do I use a computer algebra system? How do I create mathematical diagrams? How do I display my mathematical ideas on a Web page? These are tasks that undergraduate and graduate students, as well as professional mathematicians, must deal with. This book explains the use of popular mathematical tools.

This book can help you succeed in your mathematical endeavors. It is a reference that you can consult again and again as you progress in your studies. When you are a beginning math student, you may be interested in learning how to study for math tests and not so interested in learning how to do mathematical research. As time goes on, you will want to engage in advanced activities, such as computer programming, mathematical typesetting, and mathematical research. This book can accompany you as you grow mathematically.

We (the authors) have written the kind of guide that we wish had existed when we were students. We usually had to seek out a "guru" on Linux or Maple or MATLAB, or read a different book on each subject. We would have been happy to find a lot of relevant information in an easy-to-use format. Accordingly, we cover topics that we believe will be of most benefit to most mathematics students. We hope that you will be saved many hours of unnecessary toil by our advice and tutorials. We have attempted to make learning fun by including exercises and challenges that we hope will stimulate your creativity and problem-solving ability. Mathematics is not going to be easy. As Euclid said, "There is no royal road to Geometry." But at least the road can have guideposts and tourist information centers along the way.

We have written this book to help students get started in mathematics. It is our way of saying "thank you" to the mentors who helped us learn what we are now passing along to you. Besides this general acknowledgment, it is a pleasure to thank Kenneth H. Rosen for encouragement, guidance, and contribution of exercises; Robert B. Stern for support and advice; Jerrold W. Grossman and Serge G. Kruk for reviews and suggestions; Daniel Jordan and Anthony Vazzana for input about what to cover; Linda Bindner and David Garth for proofreading; and the students in our Topics in Mathematics Education: Technology class for field testing. Finally, we wish you, the reader, the very best in your mathematical development.

Part I

The Study and Practice of Modern Mathematics

Introduction

What is it like being a mathematics student? How do you study mathematics? How do you write mathematics? How do you do mathematical research and present your findings? The following eight chapters give advice on how to be an outstanding mathematics major. The purpose of this discussion is to help you succeed in your coursework and be prepared to go on to graduate school or employment as a professional mathematician.

Learning mathematics is partly about learning to think like a mathematician. Mathematicians are interested in mathematical topics, such as geometry and algebra. Mathematicians also have a way of thinking logically about problems. As you study math, you will develop and enhance your mathematical thought processes. You may notice that concepts and problems that were once difficult for you become easier and you can explain them to others. Thus, you climb the ladder of mathematical attainment. We (the authors) hope that this book helps you achieve success in your mathematical study and practice.

Chapter 1

How to Learn Mathematics

The purpose of this chapter is to help you learn mathematics more efficiently, more thoroughly, and ultimately more enjoyably.

1.1 Why learn mathematics?

Most of us study mathematics because we are taking courses in it. But why are we enrolled in these courses? Ideally, it is because we are interested in mathematics, we are intrigued by mathematical problems, we like the lore of mathematics history (what we have heard of it so far), we appreciate the beauty of the subject, we are inspired by the possibility of applying mathematics to real-world situations, and we enjoy talking about mathematics with others, perhaps even teaching others. These are some of the reasons why people are motivated to study mathematics. Another valid reason is that there are employment opportunities in mathematics.

Given these reasons for studying mathematics, it is natural to ask: How can we study and learn mathematics most effectively? In this chapter, you will learn some basic techniques to help you increase your acquisition of knowledge, improve your course grades, and enhance your overall mastery of mathematics.

In our experience, students can dramatically improve their success in mathematics studies by following the pointers in this chapter. It takes practice and discipline, but you can do it!

Please see [50] for a good discussion of effective mathematics study. A good resource about beginning college-level mathematics studies is [24]. Some good textbooks on mathematical "foundations" are [51] and [46].

1.2   Studying mathematics

The key to studying any subject is to continually apply yourself, day by day. In mathematics, as in many other academic subjects, instructors build on the previous lessons in order to go on to the next material. **It is imperative that you keep current with reading and homework assignments.**

What are some other principles of good study practice in mathematics? Mathematics is an active endeavor. You can't learn it by watching it, listening to it, or reading it. You have to *do* it. Form the habit of reading with pen and paper, trying to find examples, counterexamples, even errors.

A good way to begin is by copying statements of definitions and theorems from your text, in order to scrutinize them. For example, write out the following definitions.

> Definition. A sequence $\{a_n\}$ of real numbers is *increasing* if $a_n < a_{n+1}$ for all $n \geq 1$.

> Definition. A sequence $\{a_n\}$ of real numbers is *bounded above* if there exists a real number M such that $a_n \leq M$ for all $n \geq 1$.

When you copy mathematical statements, you begin to memorize them and make them part of your mathematical thinking. **Your instructors expect you to know the definitions and theorems.**

When you encounter a new mathematical definition, think of an example that illustrates it. For instance, the sequence $\{n\}$ is increasing and the sequence $\{1/n\}$ is bounded above (by $M = 1$).

Notice that in both of the above definitions, the quantifier "for all" means that the statements must be true for all $n \geq 1$. If they fail for even one value of n, then a sequence doesn't have the stated property.

Definitions are key ingredients in theorems. Consider this theorem.

> Theorem. A sequence of real numbers that is increasing and bounded above converges.

When you encounter a new theorem, test it. Think of an instance when the theorem holds. The sequence $\{2 - (1/n)\}$ is increasing and bounded above (by $M = 2$), and it converges to 2. Also, think of an instance where the hypothesis of the theorem isn't satisfied and the conclusion of the theorem isn't true. What if the sequence isn't increasing? Does the conclusion still hold? What if the sequence isn't bounded above? Try to find a counterexample in these situations. Doing so will help you appreciate the importance of the two ingredients in the hypothesis of the theorem.

How does one remember all the definitions and theorems? The way to remember something is to apply it, and one way to apply your knowledge is by solving problems. Perhaps the most important activity in learning mathematics is problem solving. When you solve a problem, you put your understanding to a test.

Another way to learn definitions and theorems is by studying in groups. You can quiz each other on the statements of definitions and theorems. You can also check each others' homework solutions and learn from each other when you construct and write proofs.

1.3   Homework assignments and problem solving

As you progress to higher-level mathematics courses, you will find that your instructors expect you to write your homework assignments in complete sentences, with more complete explanations than you were used to giving in lower-level courses. Thus, you face two distinct but related tasks: solving the problems and writing the solutions well.

We'll talk more about mathematical writing in the next chapter, but we would like to emphasize the most important point here: **revise your work**. When you get a homework assignment, start trying to solve the problems right away. Write down your initial ideas. Then take a second look at what you've written and revise your explanations to make them easier to follow and more elegant. Plan to work in stages, solving more of the problems and revising your written work as you go along.

A good way to improve your performance on homework assignments is to pay close attention when you are attending a mathematics lecture or reading a mathematical explanation or proof. In particular, keep in mind that each step should have a HOW and a WHY. The HOW is the justification for a step. The WHY is why you are performing that particular step (the reason it leads to the overall solution or proof). You should incorporate this "how and why" mentality into your mathematical thinking and writing. Make sure that your audience (instructor or peers) knows how and why you are performing the steps that you are performing.

Problem solving is a skill (or art) best learned by practicing on examples. Let's consider a sample homework problem related to the definitions and theorem of the previous section.

Example 1.1. Problem. Let the sequence $\{a_n\}$ be defined by the recurrence formula $a_1 = 0$, and $a_n = \sqrt{a_{n-1} + 2}$, for $n \geq 2$. Prove that $\{a_n\}$ converges.

We use a calculator or computer to find approximate values for the first few terms of the sequence:

$$0, \ 1.4142, \ 1.8477, \ 1.9615, \ 1.9903, \ 1.9975, \ 1.999397637.$$

The idea of generating **data** and looking for patterns is very important in mathematical problem solving. Based on our data, it is natural to conjecture that the sequence is increasing and bounded above by 2. Moreover, the sequence apparently converges to 2.

If we are to use the theorem of the preceding section, then we must prove that the sequence $\{a_n\}$ is increasing and bounded above. How can we prove that the sequence is increasing? Remember the definition of an increasing sequence. A sequence $\{a_n\}$ is increasing if $a_n < a_{n+1}$ for all $n \geq 1$. By the definition of $\{a_n\}$, this inequality is equivalent to the inequality

$$\sqrt{a_{n-1} + 2} < \sqrt{a_n + 2}.$$

Simplifying terms, this becomes $a_{n-1} < a_n$, which is the same as the statement to be proved, but with smaller values of the index. This gives us the idea for a proof by mathematical induction.

Can you complete a proof by mathematical induction that the sequence $\{a_n\}$ is increasing? Can you also give a proof by mathematical induction that the sequence is bounded above by 2?

If you prove these two assertions, then it will follow by the theorem that the sequence $\{a_n\}$ converges. Furthermore, we can show that the sequence converges to 2 by taking the limit of both sides of the recurrence relation. Suppose that $\lim_{n \to \infty} a_n = L$. Then

$$\lim_{n \to \infty} a_n = \lim_{n \to \infty} \sqrt{a_{n-1} + 2}$$

$$L = \sqrt{\lim_{n \to \infty} (a_{n-1} + 2)}$$

$$L = \sqrt{\lim_{n \to \infty} a_{n-1} + 2}$$

$$L = \sqrt{L + 2}.$$

Solving this equation, we find that $L = 2$.

Let's consider two more definitions and another sample problem.

Definition. A *limit point* of a set of real numbers is a real number to which a sequence of other elements of the set converges.

Definition. A set of real numbers is *closed* if it contains its limit points.

Notice the implicit quantifier *all* in the second definition. To be closed, a set must contain *all* its limit points. If there is even one limit point that it doesn't contain, then the set isn't closed. Furthermore, the definition doesn't require the set to actually have any limit points in order to be closed, only to contain them if there are any. For more on the logic of mathematical statements, a good resource is [47].

Example 1.2. Here is a sample homework problem based on the concept of a closed set.

Problem. Given sets of real numbers A and B, define $A + B = \{a + b \mid a \in A, b \in B\}$. If A and B are closed, is $A + B$ necessarily closed?

This is a "prove-or-disprove" type of problem. The time-tested solution method is to try to find a counterexample, for if you find one then you can report that the result doesn't always hold. On the other hand, as you search for a counterexample, you might start to believe that you can't find one because the result is true. So you try to prove that the result is true. If you get stuck in the proof, you can sometimes pinpoint the reason why you are stuck and this will help you find a counterexample.

Of course, to get anywhere on this problem, you have to know the definition of a closed set.

In trying to answer the problem in the affirmative (that is, prove the result), you may get stuck if the sets A and B are unbounded. Indeed, there is a simple counterexample of this type:

$$A = \{1, 2, 3, 4, \ldots\} \quad \text{and} \quad B = \left\{-1\frac{1}{2}, -2\frac{1}{3}, -3\frac{1}{4}, -4\frac{1}{5}, \ldots\right\}.$$

The sets A and B are closed (they have no limit points), but 0 is a limit point of the set $A + B$, yet 0 is not in this set. Hence, the set $A + B$ isn't closed.

There are several different types of mathematical proofs, e.g., proof by mathematical induction and proof by contradiction. When studying mathematics, you should cultivate the habit of paying attention to the kinds of proofs you learn. Try to understand their structures and you may be able to use the same structures in your own work, whether it's a homework assignment, a test, or mathematical research. More on the art and strategy of proofs can be found in [57].

Let's take a look at another problem.

Example 1.3. Problem. Find a formula for

$$1^3 - 3^3 + 5^3 - 7^3 + \cdots + (-1)^{n+1}(2n-1)^3, \quad n \geq 1.$$

We need data to solve this problem. Let $f(n)$ be the expression we are trying to find a formula for. In this case, "formula" means an expression without a summation. Here is a table of values of $f(n)$ for small n.

n	1	2	3	4	5	6	7	8
$f(n)$	1	-26	99	-244	485	-846	1351	-2024

We see that the values of $f(n)$ alternate in sign, and we also notice that n divides $f(n)$, for each value of n. Hence, it makes sense to consider the "reduced" values $f(n)/n$, as shown below.

n	1	2	3	4	5	6	7	8
$f(n)/n$	1	-13	33	-61	97	-141	193	-253

We see that the absolute values of these numbers are all odd numbers. So let's make a table of the absolute values minus 1.

n	1	2	3	4	5	6	7	8
$\lvert f(n)/n\rvert - 1$	0	12	32	60	96	140	192	252

It's unmistakable that these numbers are divisible by 4. So let's divide them by 4.

n	1	2	3	4	5	6	7	8
$(\lvert f(n)/n\rvert - 1)/4$	0	3	8	15	24	35	48	63

A little concentration reveals the pattern of these numbers: the nth number is $n^2 - 1$. Putting all our observations together, we make a conjecture:

Conjecture: $f(n) = (-1)^{n+1} n[4(n^2 - 1) + 1], \quad n \geq 1.$

Knowing what the formula (probably) is, you can prove it by mathematical induction (see Chapter 2).

1.4   Tests

It can be scary facing a mathematics test, but preparation can make the task agreeable and even fun. The best advice we can give is to try to predict the test questions and make sure that you can answer them. Imagine the good feeling you will have if those problems are presented on the test and you have practiced the answers. Even if the same problems are not posed, you will have improved your math knowledge by studying similar problems or at least ones in the same ballpark.

You should make a special point of writing down the relevant definitions, theorems, and examples that you have covered in class. Your instructor expects that you know these, so don't disappoint her or him. Knowing the basics will help you take the next step and solve the problems based on these basics.

Finally, use all of the available time on a test. You would be surprised (like your instructors) at how many students don't do this.

1.5   Inspiration

Mathematics can be a rewarding, challenging, difficult, fascinating endeavor. We are privileged to be able to study and learn mathematics, for mathematics is one of the great creations of human minds. Mathematics is a growing body of knowledge that you can be a part of, too. Every time you learn mathematics, teach someone else, or discover something new, you are adding to the total sum of mathematical thought.

Let's conclude with some perspectives on mathematics:

- "The highest form of pure thought is in mathematics." (Plato)

- "We could use up two Eternities in learning all that is to be learned about our own world and the thousands of nations that have arisen and flourished and vanished from it. Mathematics alone would occupy me eight million years." (Mark Twain)

- "Pure mathematics is, in its way, the poetry of logical ideas." (Albert Einstein)

- "[At family celebrations] when it came time for me to blow out the candles on my birthday cake, I always wished, year after year, that [Hilbert's] Tenth Problem would be solved—not that I would solve it, but just that it would be solved. I felt that I couldn't bear to die without knowing the answer." (Julia Robinson)

- "Mathematics is like looking at a house from different angles." (Thomas Storer)

- "Pure mathematics is the world's best game. It is more absorbing than chess, more of a gamble than poker, and lasts longer than Monopoly. It's free. It can be played anywhere—Archimedes did it in a bathtub." (Richard J. Trudeau)

The authors wish to add a little advice: Work on your mathematics every day. When you're at rest, the ideas will continue to click and the next day you will go further and learn more.

Exercises

1. Copy the statements of five definitions and five theorems from one of your math textbooks. Identify the use of the defined words in the statements of the theorems. Give examples that illustrate the theorems. Show how the conclusions of the theorems don't necessarily hold if the hypotheses are not satisfied.

2. Examine the proofs of three or four mathematical theorems. What is the structure of these proofs? Identify where the hypotheses of the theorems are used in the proofs.

3. Look up quotes about mathematics or mathematicians. What are your favorites?

4. What are your "secrets" for learning mathematics? What works best for you?

5. Complete the proof outlined in Example 1.1.

6. Given points $P(-1, 1)$ and $Q(2, 4)$ on the parabola $y = x^2$, where should the point R be on the parabola (between P and Q) so that the triangle PQR has the maximum possible area?

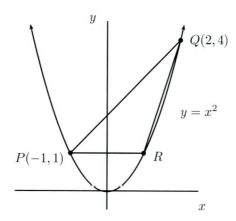

7. A license consists of two digits (0 through 9), followed by a letter (A through Z), followed by another two digits. How many different licenses are possible?

8. Find and prove a formula for

$$1^2 - 2^2 + 3^2 - 4^2 + \cdots + (-1)^{n+1} n^2, \qquad n \geq 1.$$

9. Prove the inequalities

$$2\sqrt{n+1} - 2 < \frac{1}{\sqrt{1}} + \frac{1}{\sqrt{2}} + \frac{1}{\sqrt{3}} + \cdots + \frac{1}{\sqrt{n}} < 2\sqrt{n}, \qquad n \geq 1.$$

Hint: Use integrals.

10. Which prime numbers are sums of two squares of integers? Hint: Remember to generate data.

Chapter 2

How to Write Mathematics

The purpose of this chapter is to help you write mathematical explanations and proofs. Good mathematical writing takes practice. It's also necessary to know some basic rules. Perhaps the most important feature of good mathematical writing is the revision process: writing and rewriting. This chapter discusses, with examples, the principles of mathematical writing.

2.1 What is the goal of mathematical writing?

As a mathematics student, you need to know many things in order to get started in the world of mathematics. These things include mathematical concepts, definitions, theorems, and proofs. Equally important is the knowledge of how to *write* your ideas, solutions, and proofs. Some students are surprised by this. They ask, "Why do I need to learn to write to do mathematics?" The answer is that writing is important in nearly all fields, and certainly in mathematics. You need to learn to write well so that others can follow your work.

The goal of mathematical writing is clear communication of mathematical ideas. Mathematical writing is accurate, precise, and concise. In general, writing is a skill that should be worked on and can be improved with practice. Furthermore, writing about a subject goes hand-in-hand with learning about the subject.

2.2   General principles of mathematical writing

Here we give an overview of the principles of mathematical writing. We will cover these principles in more detail in the later sections. Good resources on mathematical writing are [25] and [31].

Remember to practice the three most important principles that apply to all types of writing:

- Say something worthwhile.

- Proofread.

- Revise.

Mathematical writing has some further requirements:

- Write complete sentences. Writing complete sentences helps you to organize your thoughts and convey what you want to convey in the clearest way.

- Write accurately, precisely, and concisely. Don't write opinions, meaningless examples, or extraneous expressions. Avoid using the words "you" and "I" in math proofs.

- Avoid overly wordy and overly symbolic writing. Use a balance of words and symbols.

- Use mathematical terms and expressions properly.

- In long solutions or proofs, tell the reader in advance what you are trying to accomplish.

- Pay attention to all aspects of your writing: punctuation, spelling, mathematical content, readability, etc.

Keep your audience in mind, and write so that it is a pleasure for your audience to read your work.

2.3   Writing mathematical sentences

The basic unit of mathematical writing is the sentence. A good mathematical argument consists of sentences arranged in paragraphs that prove a theorem or give an example or illustrate how to solve a problem. Whatever the specific purpose of the mathematics, the goal is to create readable, understandable sentences. Remember that someone (usually other than yourself) must read your work. Try to put yourself in your reader's shoes and make that person's job as easy as possible by being as clear as possible.

Mathematical sentences are grammatical sentences, with the mathematics inserted in a logical way.

Example. Let $S = \{1, 2, 3, 4\}$. There are 16 subsets of S, including S itself and the empty set $\emptyset$.

Notice that the above example contains two complete sentences. (The verb in the first sentence is "$=$", which stands for "equals.") Each sentence is a blend of words and symbols.

2.4 Avoiding errors

Here are some examples of flawed and better mathematical writing:

1. Begin each sentence with a capital letter and end it with a period.

   FLAWED: let $x = 5$, then $x + 10 = 15$

   BETTER: Let $x = 5$. Then $x + 10 = 15$.

2. Begin sentences with words, not symbols.

   FLAWED: x is a positive real number, so x has a real square root.

   BETTER: Since x is a positive real number, x has a real square root.

3. Adopt a reader-friendly notation.

   FLAWED: Let x be a function from $\mathbf{R}^2$ to $\mathbf{R}^2$. Let X be the set of all such functions.

   BETTER: Let f be a function from $\mathbf{R}^2$ to $\mathbf{R}^2$. Let S be the set of all such functions.

4. Introduce terms before you use them. Don't use terms that you haven't defined.

   FLAWED: Let $x \geq 3$. Then y is a positive real number, where $y = \sqrt{x}$.

   BETTER: Let $x \geq 3$. Then $y = \sqrt{x}$ is a positive real number.

5. Use a balance of words and symbols.

   FLAWED: We claim that if we add x and y we get a number greater than or equal to 12.

   BETTER: We claim that $x + y \geq 12$.

6. Use math symbols within math expressions, not as connectives between math expressions and words.

   FLAWED: If n is > 5, then n^2 is > 25.

   BETTER: If $n > 5$, then $n^2 > 25$.

7. Write words out fully, without abbreviations.

   FLAWED: We have shown that $z = \pm i$ iff $z^2 = -1$.

   BETTER: We have shown that $z = \pm i$ if and only if $z^2 = -1$.

8. Write in paragraph style, not tabular style.

   FLAWED: Let $x = \pi$.

   Then $x > 3$.

   So x is certainly greater than e.

   BETTER: Let $x = \pi$. Then $x > 3$, so x is certainly greater than e.

9. Display long mathematical expressions. Write short mathematical expressions "in-line." Use a balance of displayed expressions and in-line expressions.

   FLAWED: Let $F(a,b,c,d,e) = (a^2 + b^2 + c^2)/(d^2 + e^2) \cdot (a + b + c)/(d + e)$. Then F is a rational function of five variables.

   BETTER: Let
   $$F(a,b,c,d,e) = \frac{a^2 + b^2 + c^2}{d^2 + e^2} \cdot \frac{a + b + c}{d + e}.$$

   Then F is a rational function of five variables.

10. Use quantifiers ("for every" and "there exists") properly.

    FLAWED: There exists an integer k for every even integer n such that $n = 2k$.

    BETTER: For every even integer n, there exists an integer k such that $n = 2k$.

2.5   Writing mathematical solutions and proofs

In writing a problem solution, the general rule is to explain every important step of your work to the reader. This is different from "showing your work."

Example 2.1. The following steps in a solution, while showing work, don't explain the math involved.

$$2^{2x} - 2^{x+2} + 3 = 0$$
$$(2^x - 1)(2^x - 3) = 0$$
$$2^x = 1, \ 3$$
$$x = 0, \ \log_2 3.$$

If you were to explain this work aloud to a friend, you would certainly add more explanation—and you would use complete sentences!

Here is a more complete explanation: We want to solve the equation

$$2^{2x} - 2^{x+2} + 3 = 0.$$

Rewriting the equation as

$$(2^x)^2 - 4 \cdot 2^x + 3 = 0,$$

we see that it is a quadratic equation in 2^x. Factoring, we obtain

$$(2^x - 1)(2^x - 3) = 0.$$

It follows that
$$2^x = 1 \quad \text{or} \quad 2^x = 3,$$
and hence
$$x = 0 \quad \text{or} \quad x = \log_2 3.$$

Notice that we write our work in sentences, we explain every important step, and we use appropriate connectives such as "it follows that" and "hence." Other useful connectives are "since" and "therefore."

A mathematical proof is a sequence of sentences that convey a mathematical argument.

Example 2.2. Let's create a proof of a simple proposition.

Proposition. Every odd integer can be expressed in exactly one of the forms $4n+1$ or $4n+3$, where n is an integer.

Here is a poor proof:

Proof. An odd integer leaves a remainder of 1 when divided by 2. Now, if we divide the integer by 4, it will leave a remainder of 1 or 3. Hence, the number is of the form $4n + 1$ or $4n + 3$. ∎

Why is this bad? There are several reasons. First, the "proof" is heavy-handed: it says that something *will* happen without explaining *why* it will happen. Second, it is confusing; it talks about "an odd integer" (in the first sentence) and "the integer" (in the second sentence). Third, the n comes out of nowhere at the end; there is no mention of n until its sudden appearance in the last sentence.

Let's try again.

Proof. An odd integer is an integer of the form $2m + 1$, where m is an integer that may be even or odd. In the first case, the original integer is of the form $4n + 1$, while in the second case it is of the form $4n + 3$. ∎

This proof started out fairly well but went astray. What went wrong? We lapsed into a words-only mode and forgot to use the notation that we set up. Good mathematical writing is a blend of words and mathematical notation. It consists of grammatically correct sentences in which some of the terms are represented by notation. Notice also how vague this proof is. The connection between m and n is unclear. A good proof explains every step clearly.

The third time is the charm.

Proof. An odd integer can be expressed in the form $2m + 1$, where m is an integer. If m is even, then $m = 2n$, for some integer n, so the given integer is of the form $2(2n)+1 = 4n+1$. If m is odd, then $m = 2n + 1$, for some integer n, so the given integer is of the form $2(2n + 1) + 1 = 4n + 3$. ∎

This is a good proof. Every step is correctly and succinctly explained, and we proved what we set out to prove. Notice that we don't belabor the obvious at the conclusion of the proof. It isn't necessary to say something like, "Since we started with an arbitrary odd integer, and we showed that it is of the form $4n + 1$ or $4n + 3$, we are done." Such a comment, while true, is tedious and useless.

Example 2.3. Let's do a proof by mathematical induction.

Problem. Prove that $5^n + 6 \cdot 7^n + 1$ is divisible by 8 for all nonnegative integers n.

The wording "for all nonnegative integers n" gives us the clue to use mathematical induction. Mathematical induction is used to prove a statement depending on an integer, for all values of the integer. A proof by mathematical induction consists of two steps: the *base case* and the *inductive step*. The base case is the statement for the least value of the variable. In our problem, since the assertion is about all nonnegative integers n, the base case is the statement for $n = 0$. Often, the base case is quite easy to prove. In our problem, the base case says that

$$5^0 + 6 \cdot 7^0 + 1$$

is divisible by 8. Since this number is equal to 8, and 8 is certainly divisible by 8, the base case is established.

The inductive step involves assuming that the assertion is true for an arbitrary value of n and showing that it must be true for the next higher value of n. Thus, we assume that $5^n + 6 \cdot 7^n + 1$ is divisible by 8 and we must show that $5^{n+1} + 6 \cdot 7^{n+1} + 1$ is divisible by 8. The assumption that the statement is true for n is called the *induction hypothesis*. Making a connection between the induction hypothesis and the statement for the next value of n is typically the crux of an induction proof.

If two expressions are both divisible by 8, then their difference is divisible by 8. So it makes sense to consider the difference between $5^n + 6 \cdot 7^n + 1$ and $5^{n+1} + 6 \cdot 7^{n+1} + 1$. We have

$$(5^{n+1} + 6 \cdot 7^{n+1} + 1) - (5^n + 6 \cdot 7^n + 1) = 5^n(5 - 1) + 6 \cdot 7^n(7 - 1)$$
$$= 4(5^n + 9 \cdot 7^n).$$

Since the $5^n + 9 \cdot 7^n$ is an even integer, the last expression is divisible by 8. This observation will allow us to establish the inductive step.

Now that we have figured out how the proof will proceed, we write our work in a neat and tidy way.

Proof. The statement is true for $n = 0$ because the expression equals 8.

Assume that $5^n + 6 \cdot 7^n + 1$ is divisible by 8, where n is a nonnegative integer. We have

$$5^{n+1} + 6 \cdot 7^{n+1} + 1 = 5^n(5 - 1) + 6 \cdot 7^n(7 - 1) + (5^n + 6 \cdot 7^n + 1)$$
$$= 4(5^n + 9 \cdot 7^n) + (5^n + 6 \cdot 7^n + 1).$$

Since $5^n + 9 \cdot 7^n$ is an even integer, $4(5^n + 9 \cdot 7^n)$ is divisible by 8. Hence, $5^{n+1} + 6 \cdot 7^{n+1} + 1$ is the sum of two integers both divisible by 8, and therefore divisible by 8.

By mathematical induction, $5^n + 6 \cdot 7^n + 1$ is divisible by 8 for all nonnegative integers n. ∎

Notice that we don't spend time in our proof talking about the logical structure of mathematical induction. We assume that the reader knows what mathematical induction is and what must be shown in this type of proof.

There are many fine books about how to write mathematical proofs, e.g., [51] and [30]. A classic book about solving math problems is [43].

2.6 Writing longer mathematical works

There are some special considerations to keep in mind when writing an article or other long project. We only touch on these issues here; a good resource is [31]. Perhaps the most important consideration is *motivation*. You should attempt to explain to your readers what you are going to do, how you are going to do it, and what the results will mean. Then show the mathematics, with relevant examples and clear statements of theorems, proofs, problems, and solutions. Finally, summarize what you have done and point out possible generalizations, limitations, and applications.

Use terms of math culture properly in your writing. A *theorem* is a main result. A *lemma* is a result whose purpose is to help prove a later theorem. A *corollary* is a result that follows easily from a theorem. You should use these terms accurately and label your results accordingly.

With a longer project, the revision process becomes even more essential. So let's talk about that now.

2.7 The revision process

Perhaps the most important element of a mathematical writing task is the revision process. Nobody writes perfect mathematical explanations the first time. We must review our work and rewrite it, striving for greater accuracy, precision, and clarity.

Read your work. After you write something, *read* it and see if it makes sense. Read it aloud. Read it to a friend. Sit quietly in a chair and read it to yourself. You will be surprised at how often what you write doesn't sound right. When this happens, rewrite your work and read it again.

When reading your work, ask yourself the following questions:

- Does each sentence make sense?

  If not, then fix the sentences that don't make sense.

- Does the overall organization make sense?

  If not, then adopt a more logical organization.

- Is every mathematical definition, theorem, and example stated correctly, using good notation?

  If not, then fix the incorrect or badly presented statements.

When you have revised your work, read it again (or have a friend read it) until you are satisfied with it. It isn't unusual to make several corrections and revisions.

Since you must revise your work, it's important that you start writing assignments early. Let's say you are taking a challenging mathematics course in which there is a weekly homework assignment consisting of five to ten problems. You should start working on the day the assignment is given, sketching your ideas about the problems, trying to find solutions, and *writing* your initial ideas. Later in the week, you should write and rewrite, while continuing to try to solve the more difficult problems. By the end of the week—after several revisions—you will have a good paper.

In fact, this kind of work regimen is necessary in graduate school, so starting the habit early is a wise move.

If you make good mathematical writing a priority, and follow the principles in this chapter, you will improve the quality of your work.

Exercises

1. Copy the statement of a theorem and its proof from one of your math textbooks. Comment on how the author(s) wrote the material. Are the ideas written in complete sentences? Is the mathematics integrated into the sentences in an appropriate way? Are all the details explained fully? Are any steps left for the reader to figure out? Are all the principles in this chapter adhered to?

2. What is wrong with the following math sentences? Correct the writing.

   Let x be a real number, then x can be rational or irrational.

   Since $\cos \theta \leq 1$, the equation $\cos^2 \theta = 5$ has no solution, for all θ.

   If x and y are both less than 2, then their product plus the sum of their squares is less than 12.

   Suppose that $\frac{a^2+b^2}{c^2+d^2} < 10$.

   From the identity $x^2 = x + 1$, we can solve for x.

   The value of n is 7. $\implies$ The value of n^2 is 49.

   Assume w.o.l.o.g. that the center of the circle is O.

   Let s be the set of prime numbers.

   Since $0 < x < y$, $x^2 < y^2$.

3. What problems can you find in the following writing? Improve the writing.

   We will prove $(x^n)' = nx^{n-1}$. By definition,

   $$(x^n)' = \lim_{y \to x} \frac{y^n - x^n}{y - x}$$
   $$= \lim_{y \to x} (y^{n-1} + y^{n-2}x + \cdots + x^{n-1})$$
   $$= nx^{n-1}.$$

4. Find an example of what you consider bad mathematical writing in a book or article and explain why it is bad.

5. In more advanced mathematics books and research articles, many details that might be explained in lower-level books and expository articles are left out. Look at some examples of exposition in your textbooks, in expository articles, and in research articles. Explain why the writing in the more advanced material is terser and what techniques are used to shorten the exposition.

6. Prove: If m_1 and m_2 are two integers both divisible by 8, then $m_1 + m_2$ is divisible by 8.

7. Write a careful proof that $\sqrt{2}$ is an irrational number. What type of proof of this?

8. Write a careful proof that e is an irrational number.

9. Prove that every square of an integer is of the form $4n$ or $4n + 1$, where n is a nonnegative integer.

10. Find a formula for
$$1^2 + 3^2 + \cdots + (2n-1)^2,$$
where n is a positive integer. Write a careful proof of your formula.

11. Find a formula for
$$\prod_{k=1}^{n} \left\{ 1 - \frac{4}{(2k-1)^2} \right\}.$$
Write a careful proof of your formula.

12. Find the value of
$$\prod_{n=0}^{\infty} \left(1 + 2^{-2^n} \right).$$
Write a careful proof of your discovery.

Chapter 3

How to Research Mathematics

The purpose of this chapter is to help you begin to research mathematics. Research can be a daunting undertaking, because it tends to be less structured than coursework, but the ideas presented here will give you concrete steps to help you get started and make progress.

3.1   What is mathematical research?

Research is an act of discovery. It is also an act of creation.

Research may mean looking up information to find answers, but in the academic world research is an inquiry or investigation with the goal of discovering new facts or revising existing theory. In mathematics, research usually means the creation of new knowledge or a careful exposition of existing knowledge.

Researching mathematics is not very different from studying mathematics. The skills you develop in studying mathematics can be put to good use in researching new mathematics.

3.2   Finding a research topic

You should choose a research topic based on something that interests you. Have you taken a course that piqued your curiosity? Do you have a professor who works in an area that you'd like to discover more about? Do you have friends who are studying a topic that sounds fascinating? These situations all indicate potential areas of research for you. To find a research topic, it is useful to look at books and articles, listen to talks, attend seminars, learn what other people are working on and where they are stuck, and go to mathematics meetings to hear about new developments.

What makes a good mathematical result? Opinions vary, but some characteristics of "good mathematics" are novelty, generality, applicability, and beauty. Is your result new? Is it surprising? Does your result generalize a previously known result? Does it make connections between two different areas of mathematics? Does it solve a longstanding problem? Does it clarify an area of mathematics that wasn't well understood? Can your result be applied to another field, such as biology or physics or computer science? Does your research explain some already-known mathematics, or mathematics history, in a new light?

Your instructors probably don't expect you to produce an earth-shattering result. But a modest contribution, well defined and presented, of any of the above types would probably be well received.

3.3   General advice

The mathematician Donald J. Lewis advised his research students to always keep several irons in the fire. This means that you should work on several topics at one time, because you never know when and where you will make a breakthrough.

Other good advice comes from Fan Chung, who says, "In mathematics whatever you learn is yours and you build it up—one step at a time."

As you progress with your research project, keep the following pointers in mind:

- In mathematics, speed doesn't matter (unless you have a deadline, of course).

- Do examples related to your work to help you understand what is going on.

- Ask yourself questions. Why does a particular piece of mathematics work the way it does? How does your discovery fit with other areas of mathematics?

- Write what you know, and update your writing as you learn more.

- Explain what you know to others (instructors and students), and listen to their feedback. You understand something best when you explain it to others. Also, your audience may point out that certain things are trivial or not understandable, and they might suggest alternate proofs or other aspects of the problem that you should look at.

- **Take every opportunity to present your findings to others, either as you go along or as a final report.** You will get practice in giving a math talk that can be useful as you consider graduate school or employment, you will get helpful feedback from your audience, and you will clarify your thinking about your topic.

3.4   Taking basic steps

Successful mathematicians use a variety of strategies to discover mathematics, but some techniques bear fruit time and time again. These techniques are a central part of the mathematical thought process.

Data Generate examples.

Observations Look at your data and find patterns or trends.

Conjectures Make guesses as to what might be true, based on your observations.

Proofs Try to explain why your guesses are true. You may want to start by proving small conjectures, ones concerned with details, before launching into a proof of a big theorem.

Generalizations Once you have obtained a proof of a significant conjecture—and now you can call your result a theorem—try to generalize your ideas. Under what conditions will the same statement hold, and can you prove it?

Connections It's always enlightening and rewarding to see your piece of math as part of the tapestry of mathematics. What role does your result play? How does it relate to other areas of mathematics?

3.5   Fixing common problems

The most common problems come from not following the advice. Here is a list of what can go wrong and how to remedy it:

- You can't get any "good results." Go back to the basics of generating examples and looking for patterns. Your results don't have to be earth-shattering. Notice details and then try to prove that the patterns you observe are really true. Little by little, you can build up interesting results.

- You write up your work in a way that is too abstract, and your instructors and peers can't understand what you have found. You need a mathematical "reality check." Remember that in mathematics, the goal is not to make things difficult, but rather to explain and prove interesting and important mathematical phenomena. If your work is too complicated, you may not have arrived at the essential mathematical truth you are aiming for. Try to simplify the statements of your results, as well as the notation and proofs. Ask yourself: What is this work about? What are the examples, theorems, and applications that other mathematicians would care about? Proceed from those examples and keep polishing your work to make it both interesting and comprehensible.

- You discover things that are already known. This happens to all mathematicians. We are excited to find a "new theorem," only to find that another mathematician has been there before. Don't despair. You were able to discover something independently, and that is good. Keep using your talents and eventually you can discover something new.

  Also, you may want to turn your "rediscovery" into a historical or expository paper about the original solution to the problem or proof of the theorem.

- You are having trouble writing up your work. Again, go back to examples. Start your written work with an example, something significant but not too complicated. This will help set the stage for why your work is important and interesting. Remember to write as you go, rather than saving all the writing until the very end of the process.

3.6   Using computer resources

What are the most useful important computer resources available in researching mathematics?

Mathematics exploration software It is difficult to overestimate the importance of generating data and looking for patterns in mathematical study and research. The following software can be tremendously helpful:

- Mathematica (Chapter 12)
- Maple (Chapter 12)
- Maxima (Chapter 12)
- MATLAB (Chapter 13)
- Octave (Chapter 13)

Mathematics typesetting software The mathematical typesetting software of choice today is LaTeX (Chapter 9). There are several good ways to access this software:

- PCTeX (www.pctex.com) is a company that supplies a particularly easy-to-use implementation of LaTeX.
- The MiKTeX project (www.miktex.org) is a complete and up-to-date implementation of TeX, including LaTeX.
- TeX Live (www.tug.org/texlive/) provides a comprehensive TeX system for all major computer platforms.

Web resources The Web is a vast sea of information. You may find the following sites especially helpful:

- MathWorld (www.mathworld.com)
- Wikipedia (www.wikipedia.org)
- MathSciNet (www.ams.org/mathscinet/)
- JSTOR (www.jstor.org)

We will describe these resources in Chapter 7.

3.7 Practicing good mathematical judgment

One of the most difficult to explain—but nevertheless very important—aspects of mathematical research is the question of mathematical judgment or taste. You choose your area of research because it is one that interests you, or your advisor, or both. Try to make sure that your work is something that others will find interesting too. Don't be ultra-abstract in your theorizing or try to prove results that are all-comprehensive. Investigate specific examples leading to well-defined results. Always subject your work to the questions, "Can I explain this to others?" and "Would others be interested?"

The issue of mathematical judgment pervades the research process and should be kept in mind at every step. An opportunity to exercise good judgment is in choosing natural notation and natural ways to formulate and present your work. This is a good reason for writing your work and presenting it to others as you develop it—it helps you keep your efforts real.

Exercises

1. Look at the William Lowell Putnam Mathematics Competitions for the past five years. You can find reports on the Competition in the *American Mathematical Monthly*. Find three questions that intrigue you and explain how to solve them.

2. Use several sources to put together a list of ten open problems in mathematics. Write a report on one of these problems, describing it in detail, its status, and the tools used to study the problem.

3. Every year student presentations are made at the annual Joint Mathematics Meetings of the American Mathematical Society (AMS) and the Mathematical Association of America (MAA). Find out what these presentations were at the most recent of these meetings. Describe the work that you find most interesting and explain why you find this work compelling.

4. Find out what kinds of projects your fellow students have done to complete a senior research requirement. Describe the work that you find most interesting and why.

5. Find out what students have accomplished in Research Experiences for Undergraduates (REUs). Describe the work that you find most interesting and why.

6. Look up the Mathematics Subject Classification. What is the purpose of the MSC? What is the numerical code for linear equations? What area does 91A46 stand for?

7. The mathematician Terrence Tao, a winner of the Fields Medal, keeps a blog of his mathematical observations and work. Take a look at some of the recent postings on his blog and describe some of the observations and questions he raises. Explain how his blog helps foster mathematical research.

8. What were the 23 open problems described by David Hilbert in 1900? What is the status of these problems (which ones have been solved or partially solved)? Why was this list of problems so important in twentieth century mathematics? Pick one of the solved problems and describe its history and how it was solved.

9. What are the Millennium Prize Problems that were announced in 2000? What is their current status? Why is this list of problems important?

Chapter 4

How to Present Mathematics

One of the joys of learning or discovering mathematics is sharing your findings with others. The purpose of this chapter is to help you give an oral presentation of mathematics that you have learned or discovered. At the end, we mention some places where you can publish your research.

4.1 Why give a presentation of mathematics?

We usually want to present mathematics in order to introduce others to what we've learned or discovered. Also, giving a mathematics talk is required in many academic programs. You should take every opportunity to present mathematics in front of others. It is excellent practice for graduate school, teaching, and job interviews. Plus, the feedback you get can help give further direction to your work.

After you give your talk at your home institution, you may have the opportunity to give it again at a mathematics meeting, such as a regional meeting, or the annual Joint Mathematics Meetings, or MathFest.

4.2   Preparing your talk

Most often, you will give a presentation based on a project that you have completed, such as a senior capstone project. Consequently, you will probably know quite a lot about your topic.

Let's suppose that you are planning to give a 30-minute talk on an aspect of mathematics that you have researched. Your audience may consist of instructors as well as peers. You should know who will make up your audience ahead of time, and you can target your talk to the right audience.

It is imperative that you practice your talk in front of others. After you finish, ask yourself what you can improve. Take suggestions from others seriously.

4.3   DOs and DON'Ts

- Practice giving your talk, ideally before an audience of peers or instructors.

- Memorize what you will say, or memorize the main points and important details, rather than read from notes. Speak slowly, calmly, and clearly. Don't speak to the board; speak directly to your audience.

- Consider starting your talk with an example, not trivial and not too difficult. Or start with a question or a quotation, something to get your audience interested.

- Don't assume that your audience members are experts in the area of your talk. In most situations, your audience members will be mathematically literate but not expert in your research area. Plan your presentation accordingly. In particular, it is important to give an example or two, and to clearly state important definitions and theorems in order to help orient your audience.

- Don't start your talk with a long list of definitions. Those who know the definitions will be bored, and those who don't know them will not be able to assimilate them. Instead, start with a meaningful example and give the definitions in context.

- If your talk is related to other problems that your audience might know about or should learn about, such as famous problems (solved or unsolved), take the time to make the connection.

- Just as in music, pauses are important.

- Don't spend a lot of time on trivial concepts and too little time on the more difficult and important ones.

- You have the right to lecture to an attentive audience. If any audience members are text messaging or doing something else annoying, you may ask the offenders (either directly or through the person in charge of the talk) to cease or go elsewhere.

- Take questions seriously.

- Don't exceed your time limit.

For more pointers on giving math talks, see [49]. You can find good advice about teaching mathematics in [32].

4.4   Using technology

What are the most important resources available for presenting mathematics?

Beamer This is software used to create slides with mathematical content typeset using LaTeX. You can find information about Beamer at

$$\texttt{en.wikipedia.org/wiki/Beamer\_(LaTeX)}$$

and in Chapter 11 of this book.

PowerPoint This software is widely available and is a good choice if you have lots of graphics to show. However, it isn't as good as Beamer for showing mathematical expressions.

Chalk and chalkboard It's fine to give your talk the old-fashioned way. You can often do a better job with chalk (or writing on a display screen) than with slides, because you must build up your work slowly, giving your audience a better chance to follow what you are doing. Also, you must be more selective in what you show, and this helps you and your audience focus on the essential material.

4.5   Answering questions

Usually, at the end of a talk (and sometimes during the talk), the audience members have a chance to ask questions of the speaker. You should take questions seriously and answer them to the best of your ability, or say that you don't know the answers. It is bad form to appeal to your advisor for an answer. Your advisor isn't giving the talk, you are.

Enjoy giving your presentation. This is your opportunity to share your mathematical knowledge and get others interested in it.

4.6   Publishing your research

After you give a talk or two on your research, and consider the feedback you've received and make appropriate improvements to your presentation, you may want to publish your work in a journal (either in print or online). There are several student journals, although you may also want to consider one of the professional research journals if your advisor judges that your work merits it.

Some journals that publish undergraduate research are listed at the Journal of Young Investigators Web site:

`http://www.jyi.org/resources/ugradPubs.html`

Some journals that publish undergraduate research are

Pi Mu Epsilon Journal

`http://www.pme-math.org`

The Pentagon (the journal of Kappa Mu Epsilon (KME) mathematical honor society)

`http://www.kappamuepsilon.org`

Rose–Hulman Undergraduate Mathematics Journal

`http://www.rose-hulman.edu/mathjournal/`

If you intend to publish your work, you probably need to write it using LaTeX (see Chapter 9). If you plan to include figures in your work, you may find LaTeX's `picture` environment sufficient; but if the figures are complicated, you probably need to use PSTricks (Chapter 10) or PostScript (Chapter 17), or generate your pictures with a computer algebra system (Chapters 12 and 13).

You should ask your advisor or another instructor to read your article before you submit it for publication. Even if your paper has already been approved as a senior research project, having another person proofread it and provide feedback is valuable.

Exercises

In the following exercises, you are asked to make presentations on various mathematical subjects. To make the presentations, you may want to use some of the tools described in Part II, such as LaTeX (Chapter 9) or Beamer (Chapter 11).

1. Give a presentation of two or more proofs of the Pythagorean Theorem.

2. Give a presentation showing a proof that e is an irrational number.

3. Give a presentation on the history of π.

4. Explain how the natural, rational, real, and complex number systems can be constructed.

5. Teach a lesson on the law of cosines. Give a proof of the law and examples of its use.

6. Give a talk on the golden ratio, ϕ. You may want to include a description of the continued fraction expansion of ϕ.

7. Describe three open conjectures about prime numbers and what is currently known about them.

8. Research the history of the Four Color Theorem. What are its origins? What was wrong with the original proofs? How was the theorem finally proved and what was the role of computers in the proof? Why do we believe that the proof using computers is correct?

9. Give a biography of a famous mathematician. Give an example of the mathematics that the person created.

10. Give an overview of the field of automated theorem proving via computer. What are the main tools used? What are some of the successes of this approach?

11. Explain what the greedy algorithm is and give several examples of its use in solving problems.

12. Give a presentation on the Axiom of Choice and its equivalent formulations in Set Theory.

13. Describe how the set of real numbers is a vector space over the set of rational numbers.

14. The area of bioinformatics is extremely active. Describe some of the mathematical tools that are applied to this area. Give an example of a mathematical theorem that has been proved in this area.

15. Explain what the field of Chaos Theory is. Describe its history and some of the mathematical ideas and results in this theory.

16. Give a talk about a mathematical model from [22] or [39] or another source.

17. Give a presentation on the mathematics of cryptography. You may want to include a discussion of the RSA Algorithm.

18. Explain what Game Theory is. Include some of its history and examples of games that have been solved mathematically.

19. How is mathematics used in studying the famous Rubik's cube puzzle? What results have been proved about the puzzle?

20. What are some of the mathematical questions that arise concerning the puzzle Sudoku? What kinds of results have been established?

21. Give a talk on a "Proof from the Book" (see [4]).

22. Explain what Environmental Mathematics is. What are some of the theories and results in this branch of mathematics? (See [14].)

Chapter 5

Looking Ahead: Taking Professional Steps

If you are an undergraduate student majoring in mathematics, you may be thinking about what you will do when you graduate. Do you plan to attend a graduate program in mathematics or in a different subject? Do you wish to become a teacher of mathematics? Would you like to be employed in an industry or a government agency? Many options exist for graduates with specializations in mathematics. While you are a student, it's prudent to take steps that will lead you toward your goals.

Here are some steps that you can take to help you in your future role as a mathematics professional:

- Learn about the culture of mathematics. Mathematical culture includes knowledge of math history, awareness of famous mathematicians and what they have accomplished, familiarity with the branches of mathematics, and exposure to famous conjectures and unsolved problems.

  As a start toward learning math culture, you may want to look up the origin of the word "mathematics." You can find this in most general dictionaries.

  A great resource for the history of mathematics is The MacTutor History of Mathematics archive, mentioned in Chapter 7.

  Here are some films that portray various aspects of mathematical life:

- *Hard Problems* (about high school students participating in the Mathematical Olympiad)

- *Julia Robinson and Hilbert's Tenth Problem* (about the mathematician Julia Robinson)

- *N is a Number* (about the mathematician Paul Erdős)

- *The Proof* (about the proof of Fermat's Last Theorem)

- *The CMI Millennium Meeting* (about the Clay Mathematics Institute prize problems meeting)

- *VideoMath Festival* (math videos shown at the International Congress of Mathematicians)

- *Music of the Spheres* (an episode of the educational television series *The Ascent of Man*)

- *Donald in Mathmagic Land* (about Donald Duck's adventures in Mathmagic Land)

Math jokes are also part of the culture. Here is the shortest math joke that we know: "Let epsilon be less than zero." Why is this funny? If you don't know, ask your Advanced Calculus instructor.

- Speaking of math culture, you should take the time to investigate the cultural and artistic productions of mathematics, such as plays, films, art (origami, sculpture, etc.), architecture, and poetry. Math is big in our societal culture these days, in television programs such as *Numb3rs* and plays such as *Proof*, in poems such as Rita Dove's "Flash Cards" and JoAnne Growney's "My Dance is Mathematics," in the sculpture of Kenneth Snelson and the knitting patterns of sarah-marie belcastro and Carolyn Yackel, in the paintings of Frank Stella and Bernar Venet, and in films such as *Starship Troopers*, *Jurassic Park*, *Good Will Hunting*, and *The Wizard of Oz*.

- Read mathematical journals and magazines that publish expository articles, such as *The American Mathematical Monthly*, *Mathematics Magazine*, and *The College Mathematics Journal*. Some journals have problem solving sections where you can be challenged and send in your solutions.

- Become a student member of the Mathematical Association of America (MAA), the American Mathematical Society (AMS), Society for Industrial and Applied Mathematics (SIAM), or Kappa Mu Epsilon (KME), a mathematics honor society. The MAA organizes state and local conferences and co-sponsors the large, annual Joint Mathematics Meetings. Take every opportunity to give a mathematics talk. If you attend the Joint Mathematics Meetings, you may be able to give a talk or participate in a poster session. In a poster session, you show a poster display of your research and people ask you questions about it; there are prizes for the top presentations.

- Take part in departmental seminars that are accessible to students. You can learn a lot from your professors and peers in these settings, and you may have the opportunity to present your own work, whether it is research or exposition. When you give a talk in your department, it helps you prepare for giving talks later as a graduate student, teacher, or employee of a business or agency.

- If a guest speaker addresses your department, you should take the opportunity to attend. As well as learning new mathematics, you may find out valuable information about where to attend a graduate program or apply for a job.

- Take part in local, regional, and national problem solving contests. These activities help keep your skills sharp. The William Lowell Putnam Mathematical Competition (organized by the MAA) is an annual problem solving competition for undergraduate students in the United States and Canada. Although students work individually, three students at each institution may be designated as a team and their combined ranks represent a rank for that institution. The Mathematical Contest in Modeling (organized by the Consortium for Mathematics and Its Applications (COMAP)) is a team competition in which institutions field teams of three students who work together to solve an applied mathematics problem. Some good books on mathematical modeling are [22] and [39]. The Internet magazine *Plus* offers an issue devoted to mathematical modeling at `plus.maths.org/issue44`.

  There are many books on mathematical problem solving. Check some out at your college library. We recommend the books by master problem-solver George Pólya.

- Speaking of your college library, browse in it for books besides your textbooks that might be helpful in your mathematical studies. Remember the advice that "there is always a better explanation available." You might find a better explanation in one of these books. We recommend the expository math books by Martin Gardner and Ross Honsberger. You may want to purchase some especially useful books for your own collection.

- While you're in the library, look at original sources in important areas of mathematics. For example, you may find books containing original papers in Calculus.

- When doing homework assignments, and any other papers, pay attention to all aspects of your writing. It is important to write mathematics well, because someone has to read what you write. There is an ulterior motive too: Someday, you may wish to ask an instructor for a letter of reference. Students who write well make a positive impression and it's a pleasure to give recommendations for them.

- Become a math tutor, grader, or teaching assistant (TA). Participating in these activities will help you improve your skills in whatever subject you are responsible for, and will help improve your "people skills" as you interact with peers and instructors.

- Prepare for the Mathematics GRE (Graduate Record Examination). Doing well on the GRE can help you get into the graduate program of your choice. To get a feel for the test, check out some books that contain test-type questions. Then review your coursework in the areas that are represented in the test. You can count on Calculus and Linear Algebra to make up a large part of the Mathematics GRE, but other areas are also important, such as Discrete Mathematics and Abstract Algebra.

- Participate in a Research Experience for Undergraduates (REU). These are funded by the National Science Foundation (NSF).

- Consider enrolling in the Budapest Semesters in Mathematics (BSM) program, in Budapest, Hungary. This is an academic program for North American undergraduate students. You can learn a lot of powerful mathematics, giving your math career a boost. Hungary has produced some of the world's top mathematicians.

- Write a senior-year mathematics paper. In some departments, this is required. It is an excellent opportunity to hone your research skills and give a math talk. Consider publishing your research in one of the student journals mentioned at the end of Chapter 4.

- Find out about service learning in mathematics. (See [23].)

- Ask yourself: What activities can I participate in that will make me more knowledge-able and skilled as I go forth in my career?

Exercises

1. Make a résumé. You may want to use LaTeX (see Chapter 9) and/or put your résumé on a Web page (see Chapter 15).

2. Find out about the process for becoming a student member of the Mathematical Association of America (MAA).

3. Find out about the process for participation in an REU (Research Experience for Undergraduates).

4. Look up recent problems in the Putnam Competition. What are the techniques used to solve these problems?

5. Look up past problems in the Mathematical Contest in Modeling. What are the techniques used to solve these problems?

6. Use the The MacTutor History of Mathematics Archive to look up a biography of Carl Friedrich Gauss. What geometric construction did Gauss discover when he was 17 years old?

7. Who is the subject of JoAnne Growney's poem "My Dance is Mathematics"?

8. Read three or four poems about mathematics/mathematicians in [15]. What are your favorites and why?

9. Look at sample problems of the GRE Mathematics Subject Test. What areas of mathematics are represented?

Chapter 6

What is it Like Being a Mathematician?

There are almost as many types of mathematician as there are types of human being. Among them are technicians, there are artists, there are poets, there are dreamers, men of affairs and many more. I well remember rising from my chair after having just solved what seemed to me an interesting and difficult problem, and saying aloud to myself: "This is beautiful music!"

RICHARD RADO (1906–1989)

[Rado was a musician as well as a mathematician.]

If you are earning an undergraduate or graduate degree in mathematics, you may wonder what you can expect in terms of future employment. In this chapter, we discuss the working life of professional mathematicians. We take "professional mathematician" to mean someone who has a college degree in mathematics and is employed in a mathematical field. This definition is very wide, although not wide enough to include amateur mathematicians. The most famous amateur mathematician was Pierre de Fermat (1601–1665). With apology to Fermat, we begin by discussing the various jobs that professional mathematicians may have. Despite the large diversity of such jobs, there are common traits that professional mathematicians share; we discuss these next. Finally, we talk about what the working conditions of a professional mathematician, especially a university professor, may entail.

According to the U.S. Bureau of Labor Statistics, mathematicians held about 3000 jobs in the U.S. in 2008. However, this figure is only a small part of the employment picture for professional mathematicians. About 14,000 undergraduate mathematics degrees were earned in the U.S. in 2009. In the same year, about 5000 mathematics master's degrees and about 1400 mathematics Ph.D.s were earned. Graduates entered a large number of different fields, including actuarial work, the scientific industry, other types of applied mathematics, government institutions (e.g., the National Security Agency), the academic industry (e.g., Wolfram Research, which makes Mathematica), school teaching, community college instruction, and university academic work.

With such a diversity of employment fields, it would be surprising if all mathematicians viewed mathematics in the same way. Each mathematician has a certain perspective on mathematics. Some believe that math is the language of science. Galileo (1564–1642) wrote that "Philosophy is written in this grand book, the universe . . . . It is written in the language of mathematics." Other mathematicians view math as a formal game played with symbols. Others believe that math is a search for truth and beauty. And there are numerous other philosophies of mathematics.

Though we may have different grand views on the meaning of mathematics, there are common traits among mathematicians. Mathematicians tend to have a confidence in numbers and equations, a belief in a "correct answer." Mathematicians are motivated by a pursuit of elegant ways to formulate and solve problems. Mathematicians have internalized a "mathematical mode of thinking," in which they look at examples, observe patterns, make conjectures, and try to prove them.

The top job in *The Wall Street Journal*'s 2009 careers survey was Mathematician. The second job on the list was Actuary, another job commonly held by professional mathematicians. It appears that professional mathematics employment yields a high rate of job satisfaction. Generally speaking, mathematicians work in comfortable office environments, with access to computing facilities, libraries, and communication facilities (such as the Internet).

You are already familiar with some mathematical jobs, like the job of high school math teacher. You have directly observed some parts of a teacher's job, such as presenting and lecturing, i.e., teaching. You can infer some of the other parts. Grading homework is an important part of being a math teacher, as is the creation of good homework assignments and tests. Writing lesson plans is also a big part of the job, and it is typically more involved than students realize. There are numerous other aspects of the job, like planning parent–teacher conferences, working with parent groups (such as the parent–teacher association), and organizing and helping with activities and fund-raisers (such as book fairs and science nights). You may also be surprised to discover that teachers do ongoing professional development, which may take the form of grant writing, summer classes, workshops, or conferences.

There are other jobs you probably know little about unless you have an acquaintance in the field. For example, most people only vaguely know that actuaries do mathematical work, although you might guess more specifically that actuaries have something to do with the actuarial tables that insurance companies use. Actuaries specialize in financial math, and they often do work for insurance companies or pension plans, either on staff or as consultants. People with an interest in finance or statistics might be drawn to actuarial work. The work is often collaborative, varied, and problem based. People or companies bring problems to you, and you work with them to find answers. In short, it is the kind of work that often draws people to study mathematics in the first place.

Many mathematicians work for the government. Some of that work is civilian and some of it is military. For example, the National Security Administration is a large employer of mathematicians (although exactly how large is unknown because that information is classified). Modern encryption is highly mathematical; naturally, mathematicians are involved in its design and analysis. But there are also many related areas: signal processing, high-speed computing, statistical analysis, and others. The NSA publishes their own internal journals, where the "cutting edge" almost certainly precedes the work of the open academic field, so it can be an exciting environment to work in.

Many larger companies have research divisions that employ mathematicians for research. The most famous of these is Bell Labs, where the transistor and the CCD (the part that makes digital cameras work) were invented. IBM research, AT&T Labs, Google, Oracle, and numerous other companies employ mathematicians, sometimes for pure research and often for applied research.

It's difficult to find out just how many companies employ mathematicians, because few places post listings for the job title "mathematician," even though it is mathematicians they will ultimately hire. A major in math, or a math major together with another minor, can lead to many careers. For example, add an interest in biology to a strong math background and you could end up doing (mathematics) work in genetics. An interest in archeology or anthropology could lead to a career in quantitative archeology. An interest in computer programming could result in work in software engineering, communications, or video game design. Every field has problems to be solved, and mathematicians train to solve problems, so though the title may say something else, you will be doing the job of a mathematician.

Of course, the job that we (the authors) are most familiar with is our own, the job of a university professor.

A professor's schedule may include teaching six to fifteen class hours per week (or less, if research grants are held that allow it), holding office hours, supervising undergraduate and graduate research, writing letters of recommendation, doing academic advising, helping colleagues with their work, participating in seminars and colloquia, and serving on departmental and institution-wide committees.

If you become a professor, you will be responsible for grading student tests and homework assignments (sometimes with the help of a student grader) and returning them promptly. You must respond to student questions, in person as well as by email and telephone. You must respond to colleagues, university administrators, and others. You may be asked to chair committees in your department or college. Meanwhile, you may be asked to play a leadership role in state or national organizations, such as the Mathematical Association of America. You may be asked by mathematics journal editors to referee articles submitted by other mathematicians for possible publication. You may be asked by book publishers to review books being considered for publication or to write reviews of published books.

A professor becomes accustomed to the fact that anyone can walk through the office door at any time with a question. The questions range from the mundane to the profound. This fact of life points to perhaps the core responsibility of university professors—that they be accessible to help others. Help takes various forms. Sometimes it is in the form of advice or information. Sometimes it is heavier work, such as reading a student's or colleague's paper and making suggestions. Sometimes the work stretches our abilities to the maximum, sometimes it is routine. From time to time, requests for assistance on mathematical problems come from students and professionals outside of one's own university. The general rule is that we are here to serve, so we try to help in whatever capacity we can. Given all of these demands, your time management skills will be well exercised.

What about research? As we have indicated in Chapter 3, the typical kind of mathematical research is the discovery of new knowledge. Research is an important aspect of being a professor. We must publish new ideas, so that our field remains relevant to the world. Other researchers build on our findings, and the body of mathematics is enriched. Exposition of known results is also vital, as it allows non-specialists to learn about new developments. Contributing questions to the problems sections of journals, and solving such problems, are also important. Typically, a faculty member must produce some publications in order to be considered for tenure (a career-long position) at an academic institution. Other achievements typically necessary for tenure are high-quality teaching and service to the department and the institution.

Research is often a collaborative effort. For example, the classification of finite simple groups, completed in 1986, was the joint effort of over 100 mathematicians from about ten nations, working over 40 years, writing about 10,000 pages of journal articles. These days, one can collaborate on research with other mathematicians quite easily, using email and telephone. The available resources for doing mathematical research are astounding. With the Web and email, one can research mathematics from almost anywhere. What is required is time and commitment.

Professional mathematicians attend conferences in order to learn about new developments in their fields, and to keep in touch with colleagues and former students. Advancements in technology, while deemed by some as distancing people socially from each other, have also made it easier than ever to stay connected with people in our professional lives. This is changing the lives of academic professionals such as mathematicians. Via email, LinkedIn, Facebook, MySpace, Twitter, etc., it is feasible to maintain professional relationships to an extent unattainable a generation ago.

Being a mathematics professor sounds like a lot of work, doesn't it? It is. But the work is both rewarding and important.

One rewarding aspect of being a university professor is teaching and supervising enthusiastic and talented young people, some of whom will become professional mathematicians.

One important aspect of being a professor is holding a high standard in terms of academic integrity. As professors, we want our ideals of honesty, responsibility, humility, and curiosity to take hold in our students. The best way to convey these tenets to others is to practice them ourselves.

If you want to be a part of the mathematical world, you can be. There is a place for you. How far you take your mathematics, and in what direction, is up to you.

Exercises

1. Find out about the following professional mathematics occupations. What do people in these occupations do? How do they use mathematics?

   - Actuary
   - Industrial Mathematician
   - National Security Agency (NSA) Mathematician
   - Computer Scientist
   - Statistician
   - Bioinformaticist

2. Go to `monster.com` and do a job search for "mathematics" or "mathematician" or "mathematical."

3. Find out what companies or organizations are the largest employers of mathematicians. What sort of work do mathematicians do in their jobs for these employers?

4. Look at the SIAM report on mathematics in industry at:

   `www.siam.org/about/mii/report.php`

   What conclusions are presented in the report?

5. What is the difference between applying mathematics and applied mathematics?

6. Give an example of a problem in the areas of operations research that is solved using mathematics. What type of mathematics is used in the solution?

7. Describe the mathematics used in the field of cryptography.

8. Describe the mathematics used in the field of agricultural economy.

9. Describe the mathematics used in the field of epidemiology.

10. Describe the mathematics used in the field of space science.

11. Describe the mathematics used in the design of "expert systems."

12. Interview two or more mathematics instructors at your college or university. What are their work schedules? What proportion of their time do they spend on teaching, preparation for teaching and office hours, committee work, student advising, grant writing, supervision of student research, participation in seminars, travel to conferences, and their own research? How do they conduct their research?

Chapter 7

Guide to Web Resources

We live in one of the best times to study mathematics, because so many resources are potentially available to us. Many of these resources are on the World Wide Web. This chapter discusses several resources that many members of the mathematical community find valuable. You may also find others that serve you well.

The Mathematical Association of America

www.maa.org

The Mathematical Association of America (MAA) is an organization that supports undergraduate mathematics education in the United States. The MAA Web site provides a wealth of information on conferences, books, articles, and research opportunities.

The MAA publishes several journals that are accessible to undergraduate students:

Math Horizons

MAA Focus

The College Mathematics Journal

Mathematics Magazine

The American Mathematical Monthly

These publications are listed in order from beginner to advanced. All the publications discuss math teaching, math problem solving, and math conferences.

The MAA also has online columns, such as "Devlin's Angle," by Keith Devlin, "The Mathematical Tourist," by Ivars Peterson, and "Launchings," by David Bressoud.

The American Mathematical Society

www.ams.org

The American Mathematical Society (AMS) is a society dedicated to professional mathematicians and graduate students in the United States. Its Web site has an abundance of information on conferences, books, articles, and research opportunities. The AMS provides MathSciNet (see below).

The AMS publishes several journals. One that is accessible to undergraduate students is

Notices of the American Mathematical Society

This journal contains expository mathematics articles, notices of math conferences, biographies of famous mathematicians, and reviews of plays, books, and films with mathematical content.

Society for Industrial and Applied Mathematics

www.siam.org

The Society for Industrial and Applied Mathematics (SIAM) is an organization dedicated to fostering "interactions between mathematics and other scientific and technological communities through membership activities, publication of journals and books, and conferences."

An online resource of SIAM that is particularly accessible to undergraduate students is the Problems and Solutions page, where mathematical problems (many from science and industry) are posed and solved.

Wikipedia

www.wikipedia.org

Wikipedia is a global, multilingual, open content encyclopedia. The content is generated by users. Wikipedia was started in 2001 by Jimmy Wales and Larry Sanger. It covers just about every topic known to humans, and its mathematical coverage is sensibly organized, easy to understand, and extensive. Most discussions contain links to other relevant sources.

We recommend Wikipedia as a good way to get your bearings in an unfamiliar mathematical area.

MathWorld

www.mathworld.wolfram.com

Mathworld is an adjunct of Wolfram Research, the company that makes Mathematica. Eric Weisstein created MathWorld, and has authored most of its 1300 articles. The coverage of topics in MathWorld ranges from excellent to minimal. Some topics are treated thoroughly with top-quality content, while others are stubs awaiting further development.

For the dazzling array of topics covered, and the wonderful graphics and links to programs and references, we recommend MathWorld as a place to gather information on mathematical topics.

The MacTutor History of Mathematics Archive

`www-history.mcs.st-and.ac.uk`

Studying mathematics goes hand-in-hand with studying its discoverers and its history. The MacTutor History of Mathematics Archive is an outstanding resource in this area. A particularly wonderful feature is the collection of biographies of mathematicians. You can search by mathematician's name, by historical period, or by mathematical topic.

MathSciNet

`www.ams.org/mathscinet`

MathSciNet, a service of the American Mathematical Society (AMS), is indispensable for researching articles and books in mathematics. You can search by title, author, subject area, and key word. You can read abstracts of articles and some book reviews. You can also search for articles that cite a particular article. Searches result in synopses of mathematical articles and books provided by *Mathematical Reviews*.

When you begin researching a mathematical topic, you can save a lot of time and energy by seeing what others have already done and what the major unsolved problems are. This will help you decide the direction of your research. Also, you will want to cite your predecessors in your work.

JSTOR

`www.jstor.org`

JSTOR is a not-for-profit organization that supplies entire articles from several mathematical journals. The service costs money, so you may want to see if your academic institution has a subscription. Some of the journals included are *The American Mathematical Monthly*, *Mathematics Magazine*, and *The College Mathematics Journal*.

ArXiv

`arxiv.org`

ArXiv, hosted by Cornell University, is a repository (archive) for research articles in many fields. You can find articles that haven't been published elsewhere. You can search by subject matter, date, author, or key word. Like MathSciNet and JSTOR, ArXiv is an indispensable research tool in mathematics.

The Mathematical Atlas

`www.math-atlas.org`

Do you know the difference between Differential Topology and Differential Geometry? The Mathematical Atlas (A Gateway to Modern Mathematics) provides an overview of the different parts of mathematics (based on the Mathematical Reviews classification scheme).

The On-Line Encyclopedia of Integer Sequences

`www.research.att.com/~njas/sequences`

Have you ever encountered an intriguing sequence of integers and wondered what was behind it? The On-Line Encyclopedia of Integer Sequences (OEIS) is a treasure-trove of information about integer sequences. You can enter the first few terms of a sequence that you are curious about and find possible matches for that sequence, as well as mathematical facts and references related to the matches. The site is run by the OEIS Foundation, which was set up by N. J. A. Sloane.

The Art of Problem Solving

`www.artofproblemsolving.com`

Do you love mathematical problem solving? If so, then The Art of Problem Solving could be the online location for you. It has lots of resources for people who enjoy posing, solving, and talking about math problems. There is plenty of content at the college level, including unsolved problems and forums for people to talk about math.

Exercises

Use Web resources to answer the following questions.

1. What is the process for someone to become a student member of the Mathematical Association of America (MAA)?

2. Use MacTutor to look up a biography of the mathematician Emmy Noether. What was her contribution to mathematics?

3. Use the OEIS to investigate the sequence 1, 2, 5, 14, 42, 132, 429, ... What famous sequence is this? List some of the mathematical contexts in which this sequence appears.

4. Use Wikipedia to research the E_8 lattice. Explain what you learn about this lattice.

5. What are the latest math news stories reported on the MathWorld site?

6. Look up Euclid's *The Elements* on the Project Gutenberg site. What is the culmination of this work (the final set of theorems proved)?

7. Find out about some unsolved math problems at "The Art of Problem Solving." Who posed the problems? Has there been any recent progress toward solutions?

8. Using the Mathematical Atlas, find out what the subjects Differential Topology and Differential Geometry are.

9. Use ArXiv to find articles on random networks. How would you narrow the search?

10. Use MathSciNet to look up Ivan Niven's article giving a proof that π is irrational. Find other articles that cite this article. Look up the same article in JSTOR. How many works does the article cite?

Chapter 8

A Mathematical Scavenger Hunt

This is a scavenger hunt, not a quiz. You can use books and the Web to find answers to the following questions. Remember the Web resources listed in Chapter 7. We hope that thinking about these questions will increase your awareness and enjoyment of the many facets of mathematics.

8.1 Mathematicians

1. What mathematician said, "There is no royal road to Geometry"?

2. What mathematician cried "Eureka!"? What was he excited about?

3. What mathematician wrote the following and where did he write it? "Cubum autem in duos cubos, aut quadratoquadratum in duos quadratoquadratos, et generaliter nullam in infinitum ultra quadratum potestatem in duos eiusdem nominis fas est diuidere, cuius rei demonstrationem mirabilem sane detexi. Hanc marginis exiguitas non caperet."

4. What two mathematicians invented Calculus?

5. Who said, "My dog Diamond knows some mathematics. Today he proved two theorems before lunch"?

6. What was the Königsberg Bridge Problem? Who solved the problem? What branch of mathematics did the solution inaugurate?

7. Who said, "It is impossible to be a mathematician without being a poet in soul"?

8. Who was Paul Erdős? What branches of mathematics did he create? Explain what an Erdős number of a mathematician is.

9. Who showed how to construct a regular heptadecagon (seventeen sides) with straight-edge and compass? How old was this mathematician at the time?

10. Who said, "Die ganzen Zahlen hat der liebe Gott gemacht, alles andere ist Menschen-werk"?

11. Who wrote the essay "A Mathematician's Apology"?

12. Who gave a lecture in 1900 to the International Congress of Mathematicians, describing 23 important unsolved mathematical problems? Discuss one of these problems that has since been solved, and one that is yet unsolved.

13. Who is known as the "Mother of Modern Algebra"?

14. What famous algorithm did George Dantzig invent?

15. Who said, "A mathematician is a machine for turning coffee into theorems"?

16. Who lived to age 100, researched in computational Number Theory, and wrote her last mathematical paper at age 87?

17. What mathematician/economist is the subject of the film *A Beautiful Mind*? What areas of mathematics did this person work in?

18. Who was the first female American to be a member of the U.S. International Mathematical Olympiad team?

19. Who were the first three women to be among the top five winners of the William Lowell Putnam Mathematical Competition?

8.2   Mathematical concepts

1. What is the Fundamental Theorem of Arithmetic?

2. What is the Fundamental Theorem of Algebra?

3. Where is the construction of an equilateral triangle given in Euclid's *The Elements*? Describe the construction. Where is the proof of the Pythagorean Theorem? Describe the proof.

4. What are the five Platonic solids? How many vertices, edges, and faces does each have?

5. What famous formula relates five important mathematical constants? Who discovered this formula?

6. What is a Mersenne prime? What is the largest one known?

7. What is the Sierpinski triangle? What is its dimension?

8. What is the Cantor Set?

9. What is the Continuum Hypothesis?

10. What is Russell's Paradox?

11. What is Ceva's Theorem?

12. What is Fermat's Little Theorem?

13. What is Fermat's Last Theorem? Who proved Fermat's Last Theorem?

14. What is Pick's Theorem?

15. What is Varignon's Theorem?

16. What is Morley's Theorem?

17. What is the Banach–Tarski Paradox?

18. What is Helly's Theorem?

19. What is Marden's Theorem?

20. What is the Prime Number Theorem?

21. What is the Weierstrass Approximation Theorem?

22. What is Zeckendorf's Theorem?

23. What is the Four Color Theorem?

24. What is Chvátal's Art Gallery Theorem?

25. What is the Kepler Conjecture? Who proved the Kepler Conjecture?

26. What is Kolmogorov Complexity?

27. What is the Green–Tao Theorem?

28. What is the Riemann Hypothesis?

8.3 Mathematical challenges

1. What is the next term in each of the following sequences?

    (a) 0, 1, 1, 2, 3, 5, 8, 13, 21, 34, 55, 89, ...
    (b) 0, 1, 2, 2, 3, 3, 4, 4, 4, 4, 5, 5, 6, 6, 6, 6, 7, 7, ...
    (c) 1, 2, 4, 6, 16, 12, 64, 24, 36, 48, 1024, 60, ...
    (d) 1, 1, 1, 2, 1, 2, 1, 5, 2, 2, 1, 5, ...

(e) 0, 0, 0, 1, 0, 1, 1, 2, 1, 3, 2, 4, ...

2. Define the sequences $\{p_n\}$ and $\{q_n\}$ recursively as follows:

$$p_1 = 1,\ q_1 = 1,$$

$$p_n = p_{n-1} + 2q_{n-1}, \quad q_n = p_n + q_n, \quad n \geq 2.$$

To what famous number does $\{p_n/q_n\}$ converge?

3. What is noteworthy about the number 6210001000?

4. What is the smallest number greater than 1 that appears at least six times in Pascal's triangle?

5. How many ways are there to make change for a dollar? You can use any number of pennies, nickels, dimes, quarters, and half-dollars.

6. What is the smallest number of people you would need to have in a room so that the probability of at least two people sharing the same birthday is greater than 0.5? Assume that the birthdays are independently and randomly distributed among the 365 days of the year (not counting Leap Days).

7. Suppose that two fair dice are rolled. Let X be the sum of the number of spots. What is the expected value of X? What is the variance of X?

8. Given a triangle in the plane, suppose that you can reflect the triangle about any of its sides, any number of times and in any order. What are the possible triangles you can start with that will tile the plane in this manner?

9. What is the only integer $n > 1$ such that $1^2 + 2^2 + \cdots + n^2$ is a perfect square? Hint: Numbers of the form $1^2 + 2^2 + \cdots + n^2$ are called *square pyramidal numbers*.

10. True of False? There exists a function from the real numbers to the real numbers that is continuous at exactly one point.

11. True of False? There exists a real-valued continuous function that is nowhere differentiable.

12. True of False? Three-dimensional Euclidean space is a disjoint union of circles.

8.4   Mathematical culture

1. What is the origin of the word "mathematics"?

2. What is the origin of the word "tangent"?

3. What is the origin of the word "lemniscate"?

4. What is the Latin term for "proof by contradiction"?

5. What does Q.E.D. stand for?

6. What mathematical equation did William Rowan Hamilton carve into a bridge in Ireland?

7. What is the smallest positive integer equal to the sum of two cubes of positive integers in two different ways? What is a famous anecdote about this number?

8. What is the Fields Medal? Who were the first two winners, and in what year did they win?

9. What are the Millennium Prize Problems? Give specific examples.

10. When and how did the William Lowell Putnam Mathematical Competition get started?

11. Where is a mathematical theorem mentioned in the film *The Wizard of Oz*?

12. What month is Mathematics Awareness Month? What is this year's theme?

8.5 Mathematical fun

1. Explain why this joke is funny: "Let epsilon be less than zero."

2. Unscramble the following mathematicians' names.

EMACHRIDSE
NLHADORE LREUE
ULJAI RNNISBOO
AIMN ERES
ALUP DROSE
ACASI WTOENN
MAME HEEMRL
TUKR GDLEO
EPOHSI INGAMER
HURT FMUONGA
PAAIYTH
NIIAASSVR ANJRMUNAA

3. What do Emanuel Lasker, Frank Morley, Sam Loyd, Claude E. Shannon, Barbara Liskov, and Noam Elkies have in common?

4. What do Martin Gardner, Persi Diaconis, Arthur T. Benjamin, and Raymond Smullyan have in common?

5. What do Ronald Graham, Péter Frankl, and Claude E. Shannon have in common?

6. What do sarah-marie belcastro, Carolyn Yackel, and Daina Taimina have in common?

7. What do Helaman Ferguson, Kenneth Snelson, and Carlo H. Séquin have in common?

8. What do Tom Lehrer, Sasha, and Jonathan Coulton have in common?

9. In the film *Starship Troopers*, what are the three greatest contributions of humans said to be?

10. Why do mathematicians and computer scientists confuse Christmas with Halloween? Hint: Think about the dates.

11. A student of mathematical physics is walking down the street when a safe falls out of a building onto his head. Although dazed from the impact, the student claims that he was fortunate. Why?

12. Which of the following words or phrases are real mathematical terms: amoeba, billiard, brane, bubble tree, curvelet, dessin d'enfant, elliptic genus, expander, gerbe, grope, horseshoe, operad, quantum chaos, skein module, snark, strange loop, systole, syzygy, topos, toric variety, train track, tropical curve?

    Hint: The mathematical terms can be found in the "What is...?" column of the *Notices of the American Mathematical Society*.

13. In the game SET, what is the maximum possible number of cards without a set?

14. Create a math movie title! Take the title of a real movie and alter it to bring in mathematical terms, phrases, or jargon. Here are some examples:

    The Wizard of Odds

    Zero to the Power Zero Flew Over the Cuckoo's Nest

    Raiders of the Lost Arctangent.

Part II

The Tools of Modern Mathematics

Introduction

With the invention and widespread manufacture of computer technology, the selection of tools available to today's mathematicians and mathematicians-in-training is stunning. It isn't necessary to focus on supercomputers. Inexpensive hand-held calculators, word processors, Web browsers, and email have profoundly changed the way we all do work and communicate about it.

Computers and software have changed the way we write about mathematics. They have changed the way we research and explore mathematics (and solve problems). And they have changed the way we share mathematics.

We have tried to include in Part II the software tools that we have found valuable, especially the tools we wish we had learned about earlier in our careers.

We begin with software for mathematical typesetting and presentation in Chapters 9, 10, and 11. While it is possible to create beautiful mathematical documents with other software, no other program has the ubiquity of LaTeX for mathematical writing. It can be used to type up a short homework assignment, to craft a class presentation, or even to write a book (like this one).

To have something to write about, we need discoveries, and Chapters 12, 13, and 14 explore some of the larger and more popular computer algebra, numerical computation, and statistical systems that we might use to make discoveries. Most students will not learn every system in these chapters, but most students will want to become acquainted with one or two of them.

The remaining chapters are more topic focused, and you may want to pick and choose the topics that interest you. If you want to create Web pages, Chapter 15 will get you started. You may find that the dynamic drawings of Geometer's Sketchpad or GeoGebra inspire you to dive into Chapter 16. Chapter 17 covers PostScript, which you may want to pursue if you enjoyed PSTricks in Chapter 10.

Many mathematicians also find programming useful, interesting, or just plain fun. Learning a whole programming language (even getting started) is beyond the scope of this book, but we have tried to give some guidance in Chapter 18 about languages to consider for your own projects.

We wrap up Part II with a discussion of free and open source software, giving particular emphasis to Linux. As with some of the other topics, we don't anticipate that all math students will become Linux devotees. However, many technical people (mathematicians included) do become ardent supporters of free software. You may become one of these people, and may even choose to contribute back to that community someday.

Chapter 9

Getting Started with LaTeX

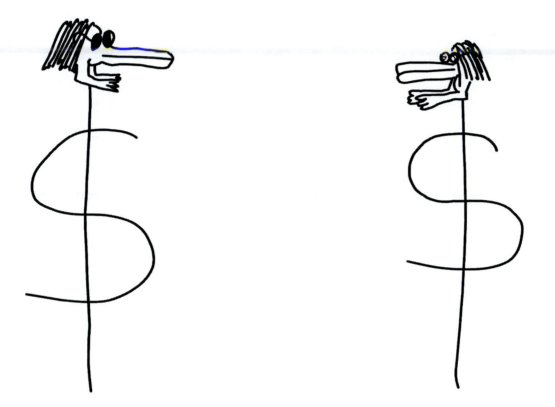

The purpose of this chapter is to help you begin using LaTeX (the most popular version of TeX), a mathematical typesetting system in which you can create, edit, typeset, and view documents containing mathematics. LaTeX is easy to learn and use. You can quickly start producing output and learn more as you go. The discussion and exercises contain many examples for you to try.

9.1 What is TeX?

TeX is a typesetting program that was invented by Donald Knuth "for the creation of beautiful books—and especially for books that contain a lot of mathematics" [29]. TeX is especially useful for typesetting mathematical documents and other documents containing complicated symbols and formatting.

TeX is a markup language. In a markup language, the user specifies how the document is to look via commands. The typesetting software interprets the commands in order to produce the pages.

Thus, TeX is both a language and a program. As a whole, it is a typesetting system.

9.2   What is LaTeX?

LaTeX is an easy-to-use version of TeX designed by Leslie Lamport. "Think of LaTeX as a house built with the lumber and nails provided by TeX" [33].

In this chapter, all examples are written in LaTeX.

> LaTeX is now extremely popular in the scientific and academic communities, and it is used extensively in industry. It has become a *lingua franca* of the scientific world; scientists send their papers electronically to colleagues around the world in the form of LaTeX input. [33]

9.3   How to create LaTeX files

LaTeX files are text files, which you should edit with simple text editors such as Notepad or Notepad++. Although they look similar, it's important to not use word processors such as Microsoft Word to create LaTeX files.

Many people use a dedicated editor for their LaTeX files. For Windows, PCTeX comes with a built-in editor, and TeXnicCenter is a popular free alternative (that works in coordination with MiKTeX). Mac users may prefer TeXShop, and Linux users might use Vi or Emacs (popular editors in the free software culture).

9.4   How to create and typeset a simple LaTeX document

A minimal LaTeX file consists of a \documentclass command, a \begin{document} command, an \end{document} command, and perhaps a line of text.

Note. Commands are preceded by the \ (backslash) symbol. Commands are case-sensitive.

Example 9.1. Here is a minimal LaTeX file.

```
\documentclass{article}
\begin{document}
This is       some         sample text.

Here is a new paragraph.
\end{document}
```

Note. The `article` document class is appropriate for many writing projects.

Note. LaTeX ignores most extra spaces in a file. A blank line (or lines) tells LaTeX to start a new paragraph.

Save the file as, say, `myfile.tex`. Then typeset the file. Programs like PCTeX, TeXnic-Center, and TeXShop will have a button that runs LaTeX to typeset your documents. Linux users will probably type a command, like `latex myfile.tex`, to do the same thing.

LaTeX will usually produce a file called `myfile.dvi` (`dvi` stands for "device independent"), although the TeXShop creates `pdf` files by default. The `dvi` or `pdf` file is displayed by a built-in viewer. Linux users will often invoke a viewer manually, perhaps with the command `xdvi myfile.dvi`.

Your displayed document should look like this:

> This is some sample text.
>
> Here is a new paragraph.

Congratulations! You have created and typeset your first document.

9.5   How to add basic information to your document

In this section you will learn how to add title, author, and date information to your documents. You will also learn how to create a numbered list (for example, a list of problem solutions in a homework paper), and you will learn how to emphasize text.

Title, author, and date commands

You can add a title, author, and date to your document with `\title`, `\author`, `\date`, and `\maketitle` commands. Try the following example.

Example 9.2. This file adds a title, author, and date to the file in Example 9.1.

```
\documentclass{article}
\title{My Document}
\author{A. Student}
\date{January 1, 2011}
\begin{document}
\maketitle
This is      some         sample text.

Here is a new paragraph.
\end{document}
```

Note. The part of the file preceding the `\begin{document}` command is called the *preamble*.

After saving the file and typesetting it, the output should look like this:

My Document

A. Student

January 1, 2011

This is some sample text.

Here is a new paragraph.

The enumerate environment

If your document is a homework assignment, you may want it to contain a numbered list of problem solutions. This can be accomplished with the `enumerate` environment.

The `enumerate` environment is opened with a `\begin{enumerate}` command and closed with an `\end{enumerate}` command. Within the environment, each item to be enumerated is preceded by an `\item` command. LaTeX will compute the item numbers for you, adjusting automatically as you add or remove items from the list.

Note. If you use several environments at once, they must be nested.

Highlighted text

Boldfaced text is produced with a `\textbf` command. Emphasized text, such as appears in a mathematical definition or a reference to a book title, is produced with an `\emph` command. The way that text is emphasized depends on context. Usually it will be italicized, but in italicized sections emphasis will in fact be non-italicized as appropriate to convey the meaning you want. Remember that LaTeX is a markup language; we tell LaTeX what we want to emphasize and it decides how to do that.

Example 9.3. This file creates a document containing a list of problem solutions.

```
\documentclass{article}
\title{My Document}
\author{A. Student}
\date{January 1, 2011}
\begin{document}
\maketitle

\begin{enumerate}

\item Here is some \textbf{boldfaced} text.

\item Here is some \emph{emphasized} text.

\end{enumerate}

\end{document}
```

After saving the file and typesetting it, the output should look like this:

My Document
A. Student
January 1, 2011

1. Here is some **boldfaced** text.

2. Here is some *emphasized* text.

Note. You can control the size of text with commands such as \Large, \large, \normalsize, \small, and \tiny. The most common way to use a size command is to place it at the beginning of a section of text enclosed in braces (which indicate grouping in LaTeX), e.g., {\Large Hello!}.

9.6   How to do elementary mathematical typesetting

In this section you will learn how to include simple mathematical expressions in your LaTeX documents. Once you are familiar with using LaTeX, you will find that you can typeset math about as easily as you can read it from left to right.

Math mode

There are several ways to put mathematical expressions into a document. The most common method is to enter *math mode* by inserting a math expression between a pair of $ signs. Thus, the input $x=3$ will produce the output $x = 3$. Observe that the font for a math x is different from the font for a normal x. It is also slightly different from the font for an italicized x.

The symbols \(and \) may be used instead of a pair of $ signs. In fact, you may prefer to use \(and \) since you may spot mistakes more quickly if you use the balanced version of the delimiters, particularly if you accidently leave one out or put an extra one in.

Centered mathematical expressions set off from the text are called "displayed" expressions, and they are generated with the \[and \] symbols. Thus, the input

$$\[y=5\]$$

produces the output

$$y = 5$$

(this is displayed).

Note. In math mode, spaces don't matter at all; they are ignored. If you want extra space, use \, (thin space), \: (medium space), or \; (thick space), commands.

Math alphabets

Greek letters are invoked by commands such as \alpha, \beta, \gamma, etc., for lowercase letters, and \Alpha, \Delta, \Theta, etc., for uppercase letters.

∅	\emptyset	⋃	\bigcup	←	\leftarrow	÷	\div
∈	\in	⋂	\bigcap	→	\rightarrow	×	\times
∉	\notin	\	\setminus	⇐	\Leftarrow	±	\pm
⊂	\subset	≈	\approx	⇒	\Rightarrow	·	\cdot
⊆	\subseteq	≐	\doteq	↔	\leftrightarrow	∘	\circ
⊃	\supset	≠	\neq	⇔	\Leftrightarrow	∗	\ast
⊇	\supseteq	≥	\geq	≅	\cong	⋯	\cdots
∪	\cup	≤	\leq	∞	\infty	…	\ldots
∩	\cap	∼	\sim	∂	\partial	′	\prime

TABLE 9.1: Mathematical symbols.

Boldfaced mathematical text is produced with a \mathbf command.

Math symbols

The following symbols are ordinary keyboard symbols and do not require special commands when used in math mode:

$$+ \quad - \quad / \quad (\ ) \quad [\ ] \quad < \quad > \quad = \quad .$$

The symbols { and } are produced with the commands \{ and \}.

Table 9.1 shows some frequently used mathematical symbols and the LaTeX commands that generate them.

Note. Certain sets are usually indicated with boldface. For example, the set of real numbers is denoted **R** (produced by the command $\mathbf{R}$).

Math functions

Many mathematical expressions are invoked by simple LaTeX commands. For example, use \sin for sin, \log for log, \ln for ln, \lim for lim, etc.

Math structures

Some useful mathematical structures are listed in Table 9.2. For example, to produce $\lim_{x \to \infty}$, use the command $\lim_{x \rightarrow \infty}$.

Example 9.4. Here is a file containing some simple mathematical expressions.

```
\documentclass{article}
\title{My Document}
\author{A. Student}
\date{January 1, 2011}
\begin{document}
\maketitle

\begin{enumerate}

\item Suppose that $x=137$.
```

Structure	Command	Example Input	Example Output
subscript	_{}	x_{10}	x_{10}
superscript	^{}	3^{20}	3^{20}
fraction	\frac{}{}	\frac{a+b}{x+y}	$\frac{a+b}{x+y}$
square root	\sqrt{}	\sqrt{x+y}	$\sqrt{x+y}$
nth root	\sqrt[n]{}	\sqrt[3]{x+y}	$\sqrt[3]{x+y}$
sum	\sum_{}^{}	\sum_{i=1}^{10}i^2	$\sum_{i=1}^{10} i^2$
product	\prod_{}^{}	\prod_{i=1}^{10}i^2	$\prod_{i=1}^{10} i^2$
integral	\int_{}^{}	\int_{0}^{\infty}x\,dx	$\int_0^\infty x\,dx$

TABLE 9.2: Mathematical structures.

```
\item Let $n=3$.   Then $n^2+1=10$.

\item The curve $y= \sqrt{x}$, where $x \geq 0$, is concave downward.

\item If $\sin \theta = 0$ and $0 \leq \theta < 2 \pi$,
then $\theta=0$ or $\theta=\pi$.

\item It is not always true that
\[\frac{a+b}{c+d}=\frac{a}{c}+\frac{b}{d}.\]

\end{enumerate}

\end{document}
```

After saving the file and typesetting it, the output should look like this:

My Document
A. Student
January 1, 2011

1. Suppose that $x = 137$.

2. Let $n = 3$. Then $n^2 + 1 = 10$.

3. The curve $y = \sqrt{x}$, where $x \geq 0$, is concave downward.

4. If $\sin \theta = 0$ and $0 \leq \theta < 2\pi$, then $\theta = 0$ or $\theta = \pi$.

5. It is not always true that
$$\frac{a+b}{c+d} = \frac{a}{c} + \frac{b}{d}.$$

Delimiter	Command	Delimiter	Command
(	\left(	)	\right)
[	\left[	]	\right]
{	\left\{	}	\right\}
|	\left| or \right|	||	\left\\| or \right\\|
⌊	\lfloor	⌋	\rfloor
⌈	\lceil	⌉	\rceil

TABLE 9.3: Delimiters.

9.7   How to do advanced mathematical typesetting

In this section you will learn how to include complex mathematical expressions in your documents.

Delimiters

Sometimes you need parentheses that are larger than normal, such as those in the expression

$$\left(\frac{a+b}{x+y}\right)^{1/3}.$$

Large delimiters (parentheses, square brackets, curly brackets, etc.) are produced with \left and \right commands. For example, the expression above is produced with the input \[\left(\frac{a+b}{x+y}\right)^{1/3}\].

Table 9.3 lists the available delimiters and the commands that produce them.

Note. Delimiters must occur in left–right pairs, but the delimiters in a pair may be of different types. If you want only one delimiter, then pair it with a period. For example, to typeset a piece-wise defined function, a left brace \left\{ would be paired with a \right., which won't show up in the display.

Arrays

Arrays (used in constructing matrices, for example) are created in the **array** environment. The **array** environment is opened with a \begin{array} command and closed with an \end{array} command. Attached to the \begin{array} command is a string of letters indicating the number of columns in the array and the type of alignment in each column. For example, the string {llrr} declares that the array contains four columns, the first and second columns aligned on the left and the third and fourth columns aligned on the right.

Line breaks (for all but the last line of the array) are indicated by two backslashes, \\. For example, to produce the matrix

$$\begin{bmatrix} a & b & c \\ d & e & f \\ g & h & i \end{bmatrix}$$

use the following commands.

```
\[
\left[
\begin{array}{ccc}
a & b & c\\
d & e & f\\
g & h & i
\end{array}
\right]
\]
```

Multi-line expressions

Multi-line expressions (e.g., in a chain of equalities) are created in the `\eqnarray*` environment. The environment is opened with a `\begin{eqnarray*}` command and closed with an `\end{eqnarray*}` command. Within the environment, each line may consist of three parts: two expressions and a relational symbol (e.g., = or ≤). The parts are separated by & symbols. As with arrays, each line (except the last) ends with a \\ command.

For example, to produce the output

$$e^x \;=\; \frac{x^0}{0!} + \frac{x^1}{1!} + \frac{x^2}{2!} + \frac{x^3}{3!} + \cdots$$

$$e^{-1} \;=\; \frac{(-1)^0}{0!} + \frac{(-1)^1}{1!} + \frac{(-1)^2}{2!} + \frac{(-1)^3}{3!} + \cdots$$

$$=\; \frac{1}{0!} - \frac{1}{1!} + \frac{1}{2!} - \frac{1}{3!} + \cdots$$

use the following commands.

```
\begin{eqnarray*}
e^x & = & \frac{x^0}{0!}+\frac{x^1}{1!}
  +\frac{x^2}{2!}+\frac{x^3}{3!}+\cdots\\
e^{-1} & = & \frac{{(-1)}^0}{0!}+\frac{{(-1)}^1}{1!}
  +\frac{{(-1)}^2}{2!}+\frac{{(-1)}^3}{3!}+\cdots\\
& = & \frac{1}{0!}-\frac{1}{1!}+\frac{1}{2!}
  -\frac{1}{3!}+\cdots
\end{eqnarray*}
```

Note. The `*` in the `eqnarray*` environment causes the equations produced to be unnumbered. If you want numbered equations, use `eqnarray` instead of `eqnarray*`.

Example 9.5. Here is a file containing several complex mathematical expressions.

Note. When typesetting long, complicated documents, it's a good idea to input the file in small blocks, typesetting at each stage. This way, if your document contains an error, you'll be able to find and correct it easily.

```
\documentclass{article}
\title{My Document}
\author{A. Student}
\date{January 1, 2011}
\begin{document}
\maketitle
```

```
\begin{enumerate}

\item Let $\mathbf{x}=(x_1,\ldots,x_n)$,
where the $x_i$ are nonnegative real numbers.
Set
\[
M_r(\mathbf{x}) = \left(\frac{x_1^r+x_2^r
+\cdots+x_n^r}{n}\right)^{1/r},
\; \; r \in \mathbf{R} \setminus \{0\},
\]
and
\[
M_0(\mathbf{x})=\left( x_1 x_2 \ldots x_n \right)^{1/n}.
\]
We call $M_r(\mathbf{x})$ the \emph{$r$th power mean}
of $\mathbf{x}$.

Claim:
\[
\lim_{r \rightarrow 0} M_r(\mathbf{x}) =
M_0(\mathbf{x}).
\]

\item Define
\[
V_n=
\left[
\begin{array}{ccccc}
1 & 1 & 1 & \ldots & 1\\
x_1 & x_2 & x_3 & \ldots & x_n\\
x_1^2 & x_2^2 & x_3^2 & \ldots & x_n^2\\
\vdots & \vdots & \vdots & \ddots & \vdots\\
x_1^{n-1} & x_2^{n-1} & x_3^{n-1} & \ldots & x_n^{n-1}
\end{array}
\right].
\]
We call $V_n$ the \emph{Vandermonde matrix} of order $n$.

Claim:
\[
\det V_n = \prod_{1 \leq i < j \leq n}(x_j-x_i).
\]

\end{enumerate}

\end{document}
```

After saving the file and typesetting it, the output should look like this:

My Document
A. Student
January 1, 2011

1. Let $\mathbf{x} = (x_1, \ldots, x_n)$, where the x_i are nonnegative real numbers. Set

$$M_r(\mathbf{x}) = \left(\frac{x_1^r + x_2^r + \cdots + x_n^r}{n} \right)^{1/r}, \quad r \in \mathbf{R} \setminus \{0\},$$

   and

$$M_0(\mathbf{x}) = (x_1 x_2 \ldots x_n)^{1/n}.$$

   We call $M_r(\mathbf{x})$ the *rth power mean* of $\mathbf{x}$.

   Claim:

$$\lim_{r \to 0} M_r(\mathbf{x}) = M_0(\mathbf{x}).$$

2. Define

$$V_n = \begin{bmatrix} 1 & 1 & 1 & \ldots & 1 \\ x_1 & x_2 & x_3 & \ldots & x_n \\ x_1^2 & x_2^2 & x_3^2 & \ldots & x_n^2 \\ \vdots & \vdots & \vdots & \ddots & \vdots \\ x_1^{n-1} & x_2^{n-1} & x_3^{n-1} & \ldots & x_n^{n-1} \end{bmatrix}.$$

   We call V_n the *Vandermonde matrix* of order n.

   Claim:

$$\det V_n = \prod_{1 \le i < j \le n} (x_j - x_i).$$

9.8 How to use graphics

There are many options for including graphics in your LaTeX documents. The simplest method is to use LaTeX's `picture` environment. For more complex pictures, you may need to use PSTricks (see Chapter 10), a package for creating images within LaTeX. Beyond that, you can include graphics created elsewhere, such as in a computer algebra system (see, for example, Chapters 12 and 13), or in applications such as Graphviz (Graph Visualization Software) and gnuplot (a graphing utility). You can also make precise mathematical images using PostScript (Chapter 17).

Example 9.6. You can use LaTeX's `picture` environment to create (somewhat limited) pictures. Here is a picture of a triangle, a square, and a circle.

```
\begin{picture}(250,75)
% draw triangle
\put(15,10){\line(1,0){50}}
\put(65,10){\line(0,1){50}}
```

```
\put(65,60){\line(-1,-1){50}}
% draw square
\put(100,10){\line(1,0){50}}
\put(150,10){\line(0,1){50}}
\put(150,60){\line(-1,0){50}}
\put(100,60){\line(0,-1){50}}
% draw circle
\put(200,35){\circle{40}}
\end{picture}
```

The commands are fairly self-explanatory. The argument (250,75) sets up space for a picture 250 × 75 units. Coordinates refer to a rectangular coordinate system with the origin at the lower-left corner. The command \line(1,0){50} produces a line of length 50 in the direction given by the vector $(1,0)$. The \circle{40} command produces a circle of diameter 40. The \put commands tell where these graphical elements are placed.

If you want to include a graphics file created outside your LaTeX document, put a \usepackage{graphicx} in your document's preamble. Use an \includegraphics command where you want an image to appear. For example, the command

```
\includegraphics[width=3in,height=3in]{5-12-13triangle.eps}
```

inserts the Encapsulated PostScript graphics file called "5-12-13triangle.eps," scaled as indicated. LaTeX can understand several types of graphics files, but PDF or PNG images generally make sense for LaTeX documents that are converted to PDF (Portable Document Format), whereas images in PS (PostScript) or EPS (Encapsulated PostScript) make sense for documents that are made into PostScript.

Example 9.7. Graphviz is a good program for quickly making graph network diagrams (directed or undirected) to use with \includegraphics. The following Graphviz input makes a directed graph depicting the squaring map on the integers modulo 10.

```
digraph {
"0" -> "0"
"1" -> "1"
"2" -> "4"
"3" -> "9"
"4" -> "6"
"5" -> "5"
"6" -> "6"
"7" -> "9"
"8" -> "4"
"9" -> "1"
}
```

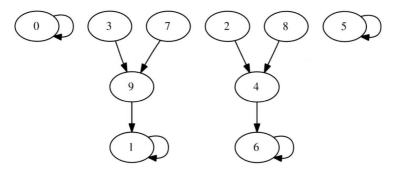

Graphviz doesn't allow the use of LaTeX commands as PSTricks does. The Graphviz Web site is at:

www.graphviz.org

For producing pictures of function plots, gnuplot is an easy-to-use tool.

Example 9.8. The gnuplot input

splot x**2-y**2

produces the saddle surface $z = x^2 - y^2$.

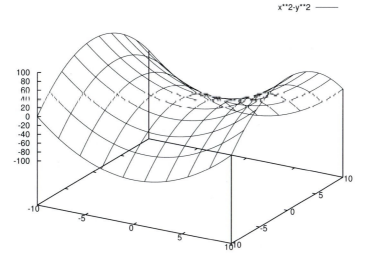

Maxima (Chapter 12) and Octave (Chapter 13) use gnuplot as their plotting engine. The gnuplot Web site is at:

www.gnuplot.info

Using PostScript, you can create precise mathematical illustrations using drawing commands within a coordinate system. See Chapter 17 for an introduction.

9.9   How to learn more

There are many aspects of LaTeX not discussed in this introduction, such as user-defined environments and packages. Here are some resources for you to investigate to learn more.

The definitive book about TeX is [29]. A good beginning book is [53].

In order to learn more about LaTeX, an excellent place to start is [33]. Other good introductory books are [12], [20], and [21]. To go further, you may want to consult [17], [19], and [18], which are all advanced books.

Some useful Web sites about TeX and LaTeX are www.ctan.org ("The Comprehensive TeX Archive Network"),

$$\text{www.tug.org/interest.html}$$

("The TeX Users Group"),

$$\text{www.emerson.emory.edu/services/latex/latex2e/latex2e\_toc.html}$$

("LaTeX Help"), and

$$\text{http://en.wikibooks.org/wiki/LaTeX}$$

(a LaTeX Wikibook).

Information about PCTeX (a user-friendly LaTeX system) is available at www.pctex.com ("PCTeX HomePage").

The project page for TeXnicCenter is www.texniccenter.org. Since TeXnicCenter is only a "front end" for LaTeX, you may want to first get MiKTeX from miktex.org. You may also want GSview for viewing PostScript files; it is available from:

$$\text{pages.cs.wisc.edu/\~ghost/gsview}$$

TeXShop is available from texshop.org and bundles a complete LaTeX system for Mac OS X.

The TeX Live project is hosted at www.tug.org/texlive. This project provides the LaTeX that is most commonly installed on Linux and Unix systems.

General references about mathematical writing and typesetting are [31], [30], and [55].

Exercises

1. What is wrong with the following LaTeX input? What is the correct way to do it?

   If m=1 and n=2, then m+n=3.

2. What is wrong with the following input? What is the right way to do it?

   If $theta = pi$, then $sin theta = 0$.

3. What is illogical about the following LaTeX input? What is a better way to do it?

   If $x=3,$ then $3x=9.$

4. Make the following equations.

$$3^3 + 4^3 + 5^3 = 6^3$$

$$\sqrt{100} = 10$$

$$(a + b)^3 = a^3 + 3a^2 b + 3ab^2 + b^3$$

$$\sum_{k=1}^{n} k = \frac{n(n+1)}{2}$$

$$\frac{\pi}{4} = \frac{1}{1} - \frac{1}{3} + \frac{1}{5} - \frac{1}{7} + \frac{1}{9} - \frac{1}{11} + \cdots$$

$$\cos\theta = \sin(90° - \theta)$$

$$e^{i\theta} = \cos\theta + i\sin\theta$$

$$\lim_{\theta \to 0} \frac{\sin\theta}{\theta} = 1$$

$$\lim_{x \to \infty} \frac{\pi(x)}{x/\log x} = 1$$

$$\int_{-\infty}^{\infty} e^{-x^2}\,dx = \sqrt{\pi}$$

5. Typeset the following sentences.

Positive numbers a, b, and c are the side lengths of a triangle if and only if $a + b > c$, $b + c > a$, and $c + a > b$.

The area of a triangle with side lengths a, b, c is given by *Heron's formula*:
$$A = \sqrt{s(s-a)(s-b)(s-c)},$$
where s is the semiperimeter $(a + b + c)/2$.

The volume of a regular tetrahedron of edge length 1 is $\sqrt{2}/12$.

The quadratic equation $ax^2 + bx + c = 0$ has roots
$$r_1, r_2 = \frac{-b \pm \sqrt{b^2 - 4ac}}{2a}.$$

The *derivative* of a function f, denoted f', is defined by

$$f'(x) = \lim_{h \to 0} \frac{f(x+h) - f(x)}{h}.$$

A real-valued function f is *convex* on an interval I if

$$f(\lambda x + (1-\lambda)y) \leq \lambda f(x) + (1-\lambda)f(y),$$

for all $x, y \in I$ and $0 \leq \lambda \leq 1$.

The general solution to the differential equation

$$y'' - 3y' + 2y = 0$$

is

$$y = C_1 e^x + C_2 e^{2x}.$$

The *Fermat number* F_n is defined as

$$F_n = 2^{2^n}, \quad n \geq 0.$$

---

6. Make the following equations. Notice the large delimiters.

$$\frac{d}{dx}\left(\frac{x}{x+1}\right) = \frac{1}{(x+1)^2}$$

$$\lim_{n \to \infty}\left(1 + \frac{1}{n}\right)^n = e$$

$$\begin{vmatrix} a & b \\ c & d \end{vmatrix} = ad - bc$$

$$R_\theta = \begin{bmatrix} \cos\theta & -\sin\theta \\ \sin\theta & \cos\theta \end{bmatrix}$$

$$\begin{vmatrix} \mathbf{i} & \mathbf{j} & \mathbf{k} \\ a_1 & a_2 & a_3 \\ b_1 & b_2 & b_3 \end{vmatrix} = \begin{vmatrix} a_2 & a_3 \\ b_2 & b_3 \end{vmatrix}\mathbf{i} - \begin{vmatrix} a_1 & a_3 \\ b_1 & b_3 \end{vmatrix}\mathbf{j} + \begin{vmatrix} a_1 & a_2 \\ b_1 & b_2 \end{vmatrix}\mathbf{k}$$

$$\begin{bmatrix} a_{11} & a_{12} \\ a_{21} & a_{22} \end{bmatrix}\begin{bmatrix} b_{11} & b_{12} \\ b_{21} & b_{22} \end{bmatrix} = \begin{bmatrix} a_{11}b_{11} + a_{12}b_{21} & a_{11}b_{12} + a_{12}b_{22} \\ a_{21}b_{11} + a_{22}b_{21} & a_{21}b_{12} + a_{22}b_{22} \end{bmatrix}$$

$$f(x) = \begin{cases} -x^2, & x < 0 \\ x^2, & 0 \leq x \leq 2 \\ 4, & x > 2 \end{cases}$$

7. Make the following multi-line equations.

$$
\begin{aligned}
1 + 2 &= 3 \\
4 + 5 + 6 &= 7 + 8 \\
9 + 10 + 11 + 12 &= 13 + 14 + 15 \\
16 + 17 + 18 + 19 + 20 &= 21 + 22 + 23 + 24 \\
25 + 26 + 27 + 28 + 29 + 30 &= 31 + 32 + 33 + 34 + 35
\end{aligned}
$$

$$
\begin{aligned}
(a+b)^2 &= (a+b)(a+b) \\
&= (a+b)a + (a+b)b \\
&= a(a+b) + b(a+b) \\
&= a^2 + ab + ba + b^2 \\
&= a^2 + ab + ab + b^2 \\
&= a^2 + 2ab + b^2
\end{aligned}
$$

$$
\begin{aligned}
\tan(\alpha + \beta + \gamma) &= \frac{\tan(\alpha + \beta) + \tan\gamma}{1 - \tan(\alpha + \beta)\tan\gamma} \\
&= \frac{\frac{\tan\alpha + \tan\beta}{1 - \tan\alpha\tan\beta} + \tan\gamma}{1 - \left(\frac{\tan\alpha + \tan\beta}{1 - \tan\alpha\tan\beta}\right)\tan\gamma} \\
&= \frac{\tan\alpha + \tan\beta + (1 - \tan\alpha\tan\beta)\tan\gamma}{1 - \tan\alpha\tan\beta - (\tan\alpha + \tan\beta)\tan\gamma} \\
&= \frac{\tan\alpha + \tan\beta + \tan\gamma - \tan\alpha\tan\beta\tan\gamma}{1 - \tan\alpha\tan\beta - \tan\alpha\tan\gamma - \tan\beta\tan\gamma}
\end{aligned}
$$

$$
\begin{aligned}
\prod_p \left(1 - \frac{1}{p^2}\right) &= \prod_p \frac{1}{1 + \frac{1}{p^2} + \frac{1}{p^4} + \cdots} \\
&= \left(\prod_p \left(1 + \frac{1}{p^2} + \frac{1}{p^4} + \cdots\right)\right)^{-1} \\
&= \left(1 + \frac{1}{2^2} + \frac{1}{3^2} + \frac{1}{4^2} + \cdots\right)^{-1} \\
&= \frac{6}{\pi^2}
\end{aligned}
$$

8. You can center text or graphics using the `center` environment. (Open the environment with `\begin{center}` and close it with `\end{center}`.) Use the `center` environment and an `\includegraphics` command to put a mathematical picture in a document. You can create the picture yourself or take it from another source. Remember to put a `\usepackage{graphicx}` command in the preamble.

9. Use LATEX's `picture` environment to make a picture of a 3–4–5 Pythagorean triangle, as below.

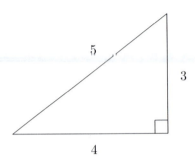

10. Add the inscribed circle of the triangle to the picture of the previous exercise.

Chapter 10

Getting Started with PSTricks

Now that you have some experience with LaTeX (see Chapter 9), you may want to spice up your LaTeX documents with some great graphics. Because of its flexibility and ease of use, the PSTricks package is a good choice for making images. The best way to learn PSTricks is by playing with examples.

10.1 What is PSTricks?

PSTricks, written by Timothy van Zandt, is a package that can be included in LaTeX documents. With PSTricks, you can use the power of PostScript's image creation language (see Chapter 17) within your LaTeX code. Many built-in commands make PSTricks easy to use.

10.2 How to make simple pictures

To use PSTricks, include a \usepackage{pstricks} command in the preamble of your LaTeX document.

Example 10.1. Let's start with a little figure displaying lines and circles.

```
\begin{pspicture}(5,5)
\psline(1,1)(5,1)(1,4)(1,1)
\pscircle[linestyle=dotted](3,2.5){2.5}
\pscircle[fillstyle=solid,fillcolor=lightgray](2,2){1}
\end{pspicture}
```

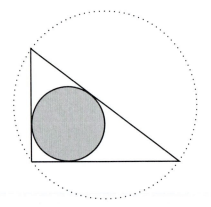

Here are the commands used to make the figure.

- The command \begin{pspicture}(5,5) starts the picture and sets aside space in a coordinate system with lower-left corner $(0,0)$ and upper-right corner $(5,5)$. The default units are equal to 1 cm.

- The command \psline(1,1)(5,1)(1,4)(1,1) draws a line path from $(1,1)$ to $(5,1)$ to $(1,4)$ to $(1,1)$, i.e., a triangle.

- The command \pscircle(3,2.5){2.5} draws a circle with center $(3, 2.5)$ and radius 2.5. Setting the parameter linestyle to dotted causes the circumference of the circle to be shown with a dotted line.

- The command \pscircle(2,2){1} draws a circle with center $(2, 2)$ and radius 1. Setting the parameters fillstyle and fillcolor to solid and lightgray, respectively, causes the interior of the circle to be shaded light gray.

- The command \end{pspicture} ends the picture.

Let's do another simple example.

Example 10.2. We draw a figure illustrating the straightedge and compass construction of an equilateral triangle.

```
\begin{pspicture}(5,5)
\psset{unit=1.5}
```

```
\psline(1,1)(3,1)(2,2.732)(1,1)
\psarc[linewidth=0.1pt](1,1){2}{0}{70}
\psarc[linewidth=0.1pt](3,1){2}{110}{180}
\end{pspicture}
```

- The command \psset{unit=1.5} changes units from 1 cm to 1.5 cm.

- The command \psarc(1,1){2}{0}{70} draws an arc of a circle of radius 2 centered at the point $(1,1)$, going from the reference angle $0°$ to the reference angle $70°$. Setting the parameter linewidth to 0.1pt defines the line width to be 0.1 points (72 points equals one inch), making the thin construction lines.

An important feature of PSTricks is that it allows the use of math mode within a picture.

Example 10.3. We draw a picture of a circle with a shaded sector, and some symbols and an equation in math mode.

```
\begin{pspicture}(4,4)
\pscircle(2,2){1.5}
\pswedge[fillstyle=solid,fillcolor=lightgray](2,2){1.5}{0}{60}
\put(2.75,1.7){$r$}
\put(2.3,2.1){$\theta$}
\put(3.25,3){$A=r\theta$}
\end{pspicture}
```

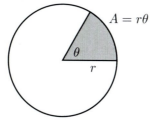

- The \put commands are from LaTeX's picture environment.

- The command \pswedge(2,2){1.5}{0}{60} creates a wedge (a sector of a circle), centered at $(2,2)$, with radius 1.5, going from reference angle $0°$ to reference angle $60°$. The parameters cause the wedge to be filled light gray.

If we want to draw a shaded sector of an ellipse, then we need to use a different method, because \pswedge only works for circles. One way to achieve the effect we want is to use a clipping path, which restricts further graphics to a specified region. Think of a clipping path as laying a window on the picture. After we have defined a clipping path, we can draw anywhere in the picture, but only the parts in the window will show.

Example 10.4. We draw an ellipse with a shaded sector.

```
\begin{pspicture}(4,4)
\psclip{\psellipse(2,2)(1.5,1)}
\psline[fillstyle=solid,fillcolor=lightgray](2,2)(4,2)(4,4)(2,2)
\endpsclip
\psellipse(2,2)(1.5,1)
\end{pspicture}
```

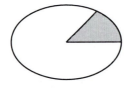

- The command \psclip sets up a clipping path defined by what is drawn in the braces that follow (the ellipse).

- The command \psellipse(2,2)(1.5,1) draws an ellipse with center $(2, 2)$, horizontal radius 1.5, and vertical radius 1.

- The command \psline draws a filled gray triangle, but only the part within the ellipse shows (because of the clipping path).

- The command \endpsclip removes the clipping path, so subsequent commands can draw anywhere.

We need to draw the ellipse a second time in order to ensure that its black line boundary is not wiped out by the gray sector.

Example 10.5. We create a picture of a parabola together with its focus and directrix. We add some labels and an equation using math mode.

```
\begin{pspicture}(-2,-2)(2,2)
\parabola{<->}(-2,1)(0,0)
\psline{<->}(-2,-1)(2,-1)
\put(0,1){\circle*{0.1}}
\put(1,0.25){\circle*{0.1}}
\put(1,-1){\circle*{0.1}}
\psline(0,1)(1,0.25)
\psline(1,0.25)(1,-1)
\put(-0.5,1){$F$}
\put(1,0.5){$P$}
```

```
\put(1,-1.5){$D$}
\put(1.5,1.5){$PF=PD$}
\end{pspicture}
```

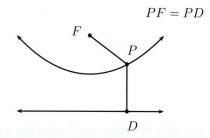

- The command `\begin{pspicture}(-2,-2)(2,2)` starts the picture and sets aside space in a coordinate system with lower-left corner $(-2, -2)$ and upper-right corner $(2, 2)$.

- The command `\parabola(-2,1)(0,0)` draws a parabola that starts at $(-2, 1)$ and has its extreme value (a minimum) at $(0, 0)$.

- Including the parameter `{<->}` in the `\parabola` and `\psline` commands puts arrows at both ends of these curves.

- The `\circle*` commands draw filled circles that represent points in the picture.

It's sometimes handy to define a drawing procedure and use it several times. We do this with a `\def` command.

Example 10.6. We make a tessellation pattern of crosses.

```
\begin{pspicture}(5,4)
\psset{unit=0.3}
\def\cross
{\psline[fillstyle=solid,fillcolor=gray]%
  (0,0)(1,0)(1,1)(2,1)(2,2)(1,2)(1,3)(0,3)%
  (0,2)(-1,2)(-1,1)(0,1)(0,0)}
\put(3,1){\cross}
\put(6,2){\cross}
\put(9,3){\cross}
\put(12,4){\cross}
\put(2,4){\cross}
\put(5,5){\cross}
\put(8,6){\cross}
\put(11,7){\cross}
\put(1,7){\cross}
\put(4,8){\cross}
\put(7,9){\cross}
\put(10,10){\cross}
\end{pspicture}
```

- The \def command defines the procedure \cross. Everything in the immediately following braces will run each time we use the \cross command.

- The percentage sign (%) at the end of a line tells PSTricks that more of the declaration of a command is coming on the next line.

Note. You may wonder whether the process of drawing elements repeatedly (e.g., the crosses in the previous example) can be done in a loop. Indeed, loops are possible in PSTricks, but in our experience, figures requiring programming constructs (loops, arrays, etc.) are created more easily using PostScript (Chapter 17).

10.3   How to plot functions

To plot a function using PSTricks, we need the pst-plot package. The examples in this section assume that we have included a \usepackage{pst-plot} command in the preamble of our document.

Example 10.7. We graph the wildly oscillating function $y = \sin(1/x)$.

```
\begin{pspicture}(-0.25,-4.25)(7.5,4.25)
\psset{xunit=3cm,yunit=3cm}
\psaxes{->}(0,0)(0,-1.25)(2.25,1.25)
\psplot[plotpoints=2500]{0.025}{2}{1 x div RadtoDeg sin}
\put(7,-0.5){$x$}
\put(-0.5,4){$y$}
\put(3,3){$y=\sin\dfrac{1}{x}$}
\end{pspicture}
```

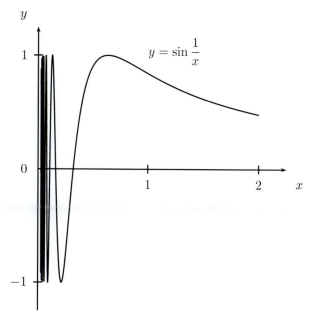

- The command `\psset{xunit=3cm,yunit=3cm}` defines units for the x and y axes.

- The `\psaxes` command creates the coordinate axes. The numbers

  `(0,0)(0,-1.25)(2.25,1.25)`

  determine that the coordinate axes intersect at $(0,0)$, the lower-left corner of the coordinate system is $(0,-1.25)$, and the upper-right corner is $(2.25,1.25)$.

- The command `\psplot` graphs the function. Setting `plotpoints=2500` specifies the number of points in the graph. The variable x goes from 0.025 to 2. (We can't let $x - 0$ because of the nature of the function.)

- Functions are defined in postfix notation, i.e., with arithmetic operators coming after the numbers they operate on rather than between them (see Chapter 17). The function definition `1 x div RadtoDeg sin` first computes $1/x$ (notice the `div` comes after the 1 and the x), then converts this real number to a degree measure, and finally computes the sine of the result.

Example 10.8. You can plot several curves together. We show a square root function and a squaring function. The square root function is depicted as a dotted curve and the square function as a dashed curve.

```
\begin{pspicture}(-0.5,-0.5)(5,5.5)
\psset{xunit=4cm,yunit=4cm}
\psaxes{->}(0,0)(1.25,1.25)
\psset{plotpoints=500}
\psplot[linestyle=dotted]{0}{1}{x sqrt}
\psplot[linestyle=dashed]{0}{1}{x 2 exp}
\put(5,-0.5){$x$}
\put(-0.5,5){$y$}
```

```
\put(1.75,3.75){$y=\sqrt{x}$}
\put(3,1.5){$y=x^2$}
\end{pspicture}
```

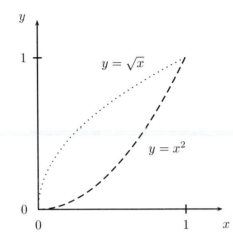

- The \psset command specifies the number of points plotted in both functions. Notice that we don't need as many points as in the previous example because these functions don't oscillate.

- In this example and the previous one, we use x as the independent variable. The psplot command expects x, so we can't use any variable that we dream up. In the next example, we do a parametric plot, so the variable is t.

Example 10.9. We make a parametric plot of a lemniscate.

```
\begin{pspicture}(-2.5,-2.5)(2.5,2.5)
\psset{xunit=3cm,yunit=3cm}
\psaxes[ticks=none,labels=none]{<->}(0,0)(-1.5,-0.75)(1.5,0.75)
\parametricplot[plotpoints=500,arrows=->,arrowscale=1.5]{0}{360}
{t cos 1 t sin 2 exp add div
t sin t cos mul 1 t sin 2 exp add div}
\put(4.5,-0.5){$x$}
\put(-0.5,2.5){$y$}
\end{pspicture}
```

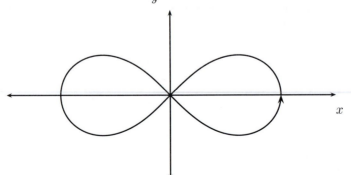

The parametric equations for the lemniscate are

$$x = \frac{\cos t}{1 + \sin^2 t}, \quad y = \frac{\sin t \cos t}{1 + \sin^2 t}, \quad 0° \leq t < 360°.$$

- In the \psaxes command, setting `ticks=none` and `labels=none` suppresses the ticks and numbers on the axes.

- The command \parametricplot sets up a parametric plot. Its parameters tell the number of points to plot, with t going from 0° to 360° and put a little arrow in the positive t direction along the curve. The x and y functions are defined in postfix notation.

10.4   How to make pictures with nodes

To make diagrams with nodes and connections, we need the `pst-node` package, so to do the examples in this section we would put a \usepackage{pst-node} command in the preamble.

Example 10.10. We make the cube graph, consisting of eight vertices and twelve edges.

```
\begin{pspicture}(-3,-3)(3,3)
\psset{radius=0.2}
% draw inner square of vertices and edges
\put(-1,1){\Circlenode{1}}
\put(1,1){\Circlenode{2}}
\put(1,-1){\Circlenode{3}}
\put(-1,-1){\Circlenode{4}}
\ncline{1}{2}
\ncline{2}{3}
\ncline{3}{4}
\ncline{4}{1}
% draw outer square of vertices and edges
\put(-2,2){\Circlenode{5}}
\put(2,2){\Circlenode{6}}
\put(2,-2){\Circlenode{7}}
\put(-2,-2){\Circlenode{8}}
\ncline{5}{6}
\ncline{6}{7}
\ncline{7}{8}
\ncline{8}{5}
% draw edges between inner and outer square
\ncline{1}{5}
\ncline{2}{6}
\ncline{3}{7}
\ncline{4}{8}
\end{pspicture}
```

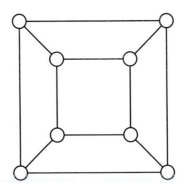

- The command \psset{radius=0.25} sets the radii of all nodes to 0.25.

- The command \Circlenode{1} defines a node in the shape of a circle, with the name "1".

- The command \ncline{1}{2} (which stands for 'node connection line') draws a line between nodes "1" and "2".

Example 10.11. We expand on the cube graph of Example 10.10 to make a diagram of a Cayley graph of the dihedral group D_4 (of order eight).

```
\begin{pspicture}(-3,-3)(3,3)
\psset{radius=0.4}
\put(-1,1){\Circlenode{E}{$e$}}
\put(1,1){\Circlenode{R}{$r$}}
\put(1,-1){\Circlenode{R2}{$r^2$}}
\put(-1,-1){\Circlenode{R3}{$r^3$}}
\ncline{->}{E}{R}
\ncline{->}{R}{R2}
\ncline{->}{R2}{R3}
\ncline{->}{R3}{E}
\put(-2,2){\Circlenode{F}{$f$}}
\put(2,2){\Circlenode{FR}{$fr$}}
\put(2,-2){\Circlenode{FR2}{$fr^2$}}
\put(-2,-2){\Circlenode{FR3}{$fr^3$}}
\ncline{->}{F}{FR3}
\ncline{->}{FR3}{FR2}
\ncline{->}{FR2}{FR}
\ncline{->}{FR}{F}
\ncline[linestyle=dashed]{E}{F}
\ncline[linestyle=dashed]{R}{FR}
\ncline[linestyle=dashed]{R2}{FR2}
\ncline[linestyle=dashed]{R3}{FR3}
\put(0.25,1.25){$r$}
\put(-1,1.7){$f$}
\end{pspicture}
```

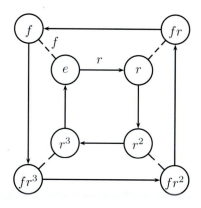

- The parameter {->} in an \ncline command puts an arrowhead on the end of the line.

Example 10.12. We make a picture of a finite state automaton that accepts binary strings with an even number of 1s.

```
\begin{pspicture}(6,3)
\psset{radius=0.5}
\put(2,1){\Circlenode[doubleline=true]{EVEN}{even}}
\put(4,1){\Circlenode{ODD}{odd}}
\ncarc{->}{EVEN}{ODD}
\ncarc{->}{ODD}{EVEN}
\nccircle{->}{EVEN}{0.5}
\nccircle{->}{ODD}{0.5}
\put(1.5,1){\voctor(1,0){0.5}}
\put(2.4,2.25){$0$}
\put(4.4,2.25){$0$}
\put(3.4,1.4){$1$}
\put(3.4,0.6){$1$}
\end{pspicture}
```

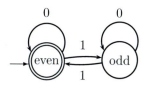

- Setting the parameter doubleline=true creates a double circle node.

- The command \ncarc draws an arc.

10.5   How to learn more

PSTricks supports the use of color. The following example depicts the five tetrominoes packed into a 3×7 box with one empty square. See Figure 1 of the color insert.

```
\begin{pspicture}(8,4)
\psline[fillstyle=solid,fillcolor=red]
(1,1)(3,1)(3,3)(1,3)(1,1)
\psline[fillstyle=solid,fillcolor=yellow]
(1,3)(5,3)(5,4)(1,4)(1,3)
\psline[fillstyle=solid,fillcolor=green]
(3,1)(5,1)(5,2)(6,2)(6,3)(4,3)(4,2)(3,2)(3,1)
\psline[fillstyle=solid,fillcolor=blue]
(5,1)(8,1)(8,2)(7,2)(7,3)(6,3)(6,2)(5,2)(5,1)
\psline[fillstyle=solid,fillcolor=lightgray]
(5,4)(8,4)(8,2)(7,2)(7,3)(5,3)(5,4)
\end{pspicture}
```

When the five tetrominoes are put in a 3×7 box, can you figure out why the empty square can't be the center square?

A good source for lots of examples, in black-and-white and color, is the PSTricks Web site at:

www.tug.org/PSTricks/main.cgi

The standard reference for LaTeX graphics is [19].

Exercises

1. Make a picture of a 5–12–13 Pythagorean triangle, as below.

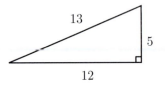

2. Make a diagram like the one that follows, illustrating Archimedes' demonstration of the formula for the area of a circle.

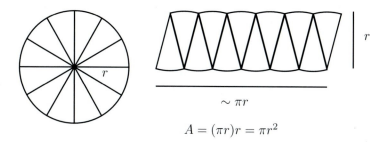

$$\sim \pi r$$

$$A = (\pi r)r = \pi r^2$$

3. Make a picture of the Venn diagram below.

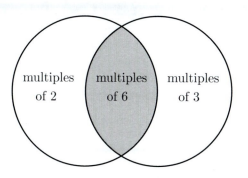

4. Make a picture of an ellipse and its two foci F_1 and F_2, illustrating the relation $F_1 P + F_2 P =$ constant, where P is a point on the ellipse.

5. Write each of these postfix expressions in standard form:

```
x 1 add 2 exp

x 1 add 2 exp 1 x sub div

x x sin mul

2 x sin x cos mul mul
```

6. Plot the function
$$f(x) = \begin{cases} x^2, & 0 \le x \le 2 \\ -x^2, & -2 \le x < 0. \end{cases}$$

7. Plot $y = \sin x$ and $y = \cos x$ on the same coordinate system, for $0 \le x \le 2\pi$. Show the sine function as a solid curve and the cosine function as a dotted curve.

8. Plot $y = \sqrt{x}\sin(1/x)$, for $0 < x \le 2$. On the same coordinate system, plot the functions $y = \sqrt{x}$ and $y = -\sqrt{x}$, for $0 \le x \le 2$, with these functions shown as dotted curves.

9. Plot the cardioid given by the parametric equations

$$x = \cos t(1 - \cos t)$$

$$y = \sin t(1 - \cos t), \quad 0 \le t \le 2\pi.$$

10. Draw a graph consisting of two sets of three nodes and all nine possible line connections between the two sets.

11. Draw a graph consisting of fives nodes and all possible line connections between them.

12. Make a diagram of the icosahedral graph below.

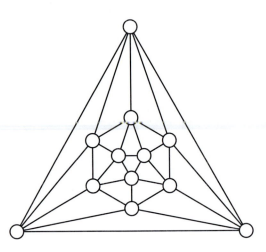

13. Make a diagram of the Petersen graph, with ten vertices and fifteen edges, as below. Color the vertices with three colors so that no connected vertices are the same color.

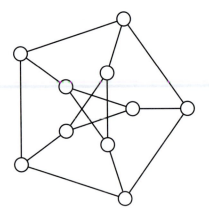

14. Make a diagram of the Cayley graph of the eight-element group $\mathbf{Z}_4 \times \mathbf{Z}_2$, with generators $(1, 0)$ and $(0, 1)$.

15. Draw a finite state automaton that accepts binary strings that do not contain two consecutive 0's. You can label the states "good string not ending in 0," "good string ending in 0," and "bad string."

Chapter 11

Getting Started with Beamer

In this chapter, we take a look at one of the most popular ways to give a math talk, using Beamer.

11.1   What is Beamer?

So, you've been asked to make a presentation on a mathematical topic. How do you project mathematical symbols and formulas for your audience to see? Beamer was designed by Till Tantau for just this purpose. Beamer is a LaTeX document class that enables you to project a slide show composed in the LaTeX mathematical typesetting system. In this chapter, we'll assume that you know the basics of LaTeX and have some material composed in LaTeX (see Chapter 9). We'll cover the essentials of Beamer.

11.2   How to think in terms of frames

Starting with your LaTeX document, Beamer produces a PDF document that you can project onto a screen. In terms of ease of use and flexibility, Beamer has much in common with Microsoft's PowerPoint. However, with Beamer you can project work composed in LaTeX. This is very convenient for math students (and professional mathematicians), because mathematical formulas and symbols typically play a large role in math presentations.

The basic unit of Beamer is a frame, which generates the contents of a single slide. (We'll learn about frames in the next section.) You will need to think about breaking up your work into frames. If you've already written a paper for an assignment, then you will want to distill the main points of your work into easily digested units to make up the frames. A little word of advice on this point: Concise is usually better than verbose. You should make each frame easy to comprehend for the slow readers in your audience. You can (and will) say more than what is written in a frame, so you have plenty of opportunities for verbal embellishment.

As you prepare your talk, keep in mind the "secrets" of good slide presentations.

The Seven Secrets of Superb Slide Shows:

- Put a small amount of material on each slide.

- Keep slides simple (no distractions).

- Keep slides consistent in style.

- Use pictures where appropriate.

- Use numbered or itemized lists where appropriate.

- Do examples.

- Show references where people can look up more information about your subject.

11.3   How to set up a Beamer document

Here are the basic elements of a beamer document. If you want to use this skeleton as a template for your work, go right ahead. You can simply change the material to make it suit your needs.

Notice that we use `beamer` (rather than `article`) as the document class in the preamble of the LaTeX document. Remember that you can use LaTeX commands for mathematical typesetting within your document. We don't do much of that in this example, but we do put $ signs around numbers.

A Sample Beamer Document

```
\documentclass{beamer}

\title{My Wonderful Topic}
\author{Myself}
\institute{My University}

\begin{document}

\begin{frame}
\titlepage
\end{frame}

\begin{frame}
\begin{abstract}
We will discuss some properties of the number $12$.
\end{abstract}
\end{frame}

\begin{frame}
\frametitle{Facts about $12$}

\begin{itemize}
\item $12$ has six positive divisors.
\item $12$ is twice a perfect number ($6$).
\item $12$ is one of the legs of a famous
      Pythagorean triangle ($5$--$12$--$13$).
\end{itemize}
\end{frame}

\begin{frame}
\frametitle{Open Questions}

\begin{itemize}
\item How many positive integers less than $100$
      have six positive divisors?
\item Is twice a perfect number always an abundant number?
\item Are there other Pythagorean triangles with $12$
      as a side length?
\end{itemize}
\end{frame}

\end{document}
```

As you can see, the basic ingredient of a beamer document is the frame. Frames generate what you and your audience see on the screen. Similar to other LATEX constructions (such as theorems and lists), frames are made with a `\begin{frame}`...`\end{frame}` construction. Our first frame (with the `\titlepage` command) produces the title page. The second frame contains an abstract for the talk. The other frames have titles given by the `\frametitle` command.

Note. After you create each new frame, rebuild your LATEX document to check that things are working the way that you want. This will help you pinpoint the locations of errors (which we all inevitably make).

You can use ordinary LATEX commands in the frames of your presentation. In the example above, we've used math mode for the numbers. In practice, the math can be quite complex (anything that LATEX can handle). But remember the general rule to keep things as simple as possible for your audience.

This is a good time to reiterate some advice about giving talks (see Chapter 4). It's often a good idea to start a talk with an example. This can be a sample calculation or an easy-to-understand example of a phenomenon that you wish to describe. Too many talks start off with a laundry list of definitions and theorems, leaving the audience members in need of life preservers. In a math talk, an example is always welcome.

11.4 How to enhance a Beamer presentation

Would a picture enhance your math talk? If you are giving examples of the Pythagorean Theorem, you would want to show a picture of a right triangle. You can include images in your Beamer presentation the same way you do in other LATEX documents. Simply put a \usepackage{graphicx} command in the preamble of your document, and use an \includegraphics command in the frame where you want the image to appear.

```
\begin{frame}
\frametitle{A famous Pythagorean triangle}

Here is a picture of the $5$--$12$--$13$ Pythagorean triangle.

\begin{center}
\includegraphics[width=3in,height=3in]{5-12-13triangle.eps}
\end{center}
\end{frame}
```

The LATEX picture environment also works in Beamer documents, and it may be suitable if your pictures are simple.

Or you can use Beamer together with PSTricks (see Chapter 10). You will need a \usepackage{pstricks} command in the preamble of your document. Use the construction

```
\begin{frame}{PSTricks}
\end{frame}
```

for frames that contain PSTricks pictures.

You can do much more with Beamer, such as show the elements of a list one point at a time. To do this, put a \pause command in an item in an enumerate environment.

```
\begin{enumerate}
\item We will pause to discuss this item.  \pause
\item We will pause to discuss this item too.  \pause
\item Now we will discuss the final item.
\end{enumerate}
```

Now that you know the basics of Beamer, we encourage you to explore. For example, you can learn how to do each of the following:

- Enhance the look of your slides with "Beamer blocks."

- Jump forward and backward in your presentation, and link to Web documents.

- Generate a Table of Contents for your talk.

11.5 How to learn more

You can learn more about Beamer at:

    ctan.org/tex-archive/macros/latex/contrib/beamer/doc/beameruserguide.pdf

There are many other good guides to Beamer on the World Wide Web. You can search for guides by entering "Beamer tutorial" in a Web browser. The LaTeX Beamer Class Homepage is located at:

    latex-beamer.sourceforge.net

You may want to create a simple Beamer presentation, perhaps using the examples in this chapter, and then add to your presentation by following some of the more detailed advice that you can find on the Web.

Exercises

Use Beamer to give presentations on the following topics

1. Give a presentation on the Pythagorean Theorem. Be sure to include diagrams of right triangles.

2. Give a presentation on Pascal's triangle.

3. Give a presentation on the Fundamental Theorem of Calculus. Include diagrams.

4. Give a presentation on Linear Programming optimization methods. You may want to include a discussion of the Simplex Algorithm.

5. Give a presentation on Numerical Analysis used in solving equations.

6. Give a presentation on some theorems of Graph Theory. Include diagrams.

7. Give a presentation on fast Fourier transforms.

8. Give a presentation on a mathematician chosen from [5] or [6]. Give biographical background on the person and describe some of the mathematics that the person created.

9. Give a presentation on a topic in mathematics and art, chosen from [28], [13], or [41].

Chapter 12

Getting Started with Mathematica®, Maple®, and Maxima

Arguably the most important software tool for a mathematics student is the computer algebra system (CAS). No other software puts so much mathematical potential into a single tool.

12.1   What is a computer algebra system?

A computer algebra system is a program with which you can perform calculations, evaluate functions, create graphics, and develop your own programs. The key feature of computer algebra systems is the ability to manipulate expressions symbolically. Typical manipulations possible in a CAS include simplifying expressions, factoring, taking derivatives, computing integrals (symbolically and numerically), and solving systems of equations. This chapter explores the basics of three popular computer algebra systems, and it contains simple examples for you to try.

Mathematica, created by Stephen Wolfram, is probably the world's most recognized computer algebra system. It was originally released in 1988 and is still being developed and improved. In addition to its raw power, one notable feature of Mathematica is its use of the "notebook." Mathematica notebooks allow a user to combine written text with calculations in one integrated document.

Maple, created by Waterloo Maple under the trade name Maplesoft, dates back to the early 1980s. It is one of the dominant commercial computer algebra systems and favored by many institutions. Like Mathematica, it can create integrated documents, called worksheets, that combine text, calculations, and hyperlinks.

Maxima is a free software computer algebra system. It is derived from an early computer program, Macsyma, which dates back to the 1960s and was made available under an open source license starting in 1998. Because it is free, it is especially attractive to students. Maxima can create documents that combine text with calculations through a graphical front end called wxMaxima.

12.2   How to use a CAS as a calculator

When using any computer algebra system as a calculator, it is important to understand that it is a bit difficult to translate mathematical writing directly to the computer. Humans instinctively adapt to ambiguity, but software is less flexible. For example, mathematicians use parentheses for grouping, as in $(x-1)(x+2)$. But they also use parentheses to indicate ordered pairs, like $(1,3)$, and to denote functions $f(x)$.

Another ambiguity that is perhaps a bit more subtle occurs with "equals." Humans have little trouble understanding that sometimes we intend equals to assign values, as with "let $x = 2$." At other times, we mean to assert equality; a circle is the set of points (x,y) such that $x^2 + y^2 = 1$.

Each computer algebra system addresses the job of translating mathematical syntax into unambiguous "computer syntax" in its own way. To a first-time software user this can feel somewhat unintuitive, even quirky, but mastering the language of your favorite CAS is an important part of using it effectively.

Mathematica as a calculator

Since Mathematica notebooks are used for text as well as calculations, you will almost immediately notice one idiosyncrasy when you try to use Mathematics as a calculator. The ENTER key does not run a calculation (it ends a paragraph or makes it possible to enter multi-line computations). To use Mathematica as a calculator, type the expression you wish to evaluate and press SHIFT+ENTER. (The special ENTER key on the lower-right corner of the keypad of an extended keyboard also works.)

Example 12.1. We add 2 and 2 to get ... 4.

```
In[1]:= 2 + 2
```

```
Out[1]= 4
```

Note. Mathematica assigns line numbers to the input and output, e.g., the "`In[1]:=`" and "`Out[1]=`" above. You do not type them yourself.

To multiply two numbers, type the numbers with a space between them. Use a caret (^) for exponentiation. Notice that Mathematica can handle very large numbers easily, even numbers with hundreds of digits.

Example 12.2. A product, and the value of 3^{100}.

```
In[2]:= 1024 59049
```

```
Out[2]= 60466176
```

```
In[3]:= 3^100
```

```
Out[3]= 515377520732011331036461129765621272702107522001
```

The values of important mathematical constants (such as π, e, and i) are stored in Mathematica. To distinguish them from variables you might create yourself, Mathematica's internal constants are capitalized (Pi, E, I, etc.).

The built-in constants are handled algebraically, but we can request the numerical value of an expression with the N function.

Example 12.3. Calculations with π, e, and i.

```
In[4]:= Pi
```

```
Out[4]= Pi
```

```
In[5]:= N[Pi]
```

```
Out[5]= 3.14159
```

```
In[6]:= N[E]
```

```
Out[6]= 2.71828
```

```
In[7]:= I I
```

```
Out[7]= -1
```

If you want a numerical result given to a high degree of accuracy, use the command N[_,_]. The first argument of this function is the number to be calculated. The second argument is the number of decimal places to which the number is computed.

Example 12.4. We calculate π to 100 decimal places.

```
In[8]:= N[Pi,100]
```

```
Out[8]= 3.141592653589793238462643383279502884197169399375105820974944592307816406286208998628034825342117068
```

Note. You can obtain information about a specific command by typing a question mark followed by the name of the command. For instance, to find out about the function N, type:

```
In[9]:= ? N
```

```
N[expr] gives the numerical value of expr. N[expr, n]
attempts to give a result with n-digit precision.
```

In addition to processing numerical calculations, Mathematica performs algebraic operations. If a variable has not been assigned a value, Mathematica will work with it algebraically.

Example 12.5. We set a equal to 17, and then calculate with a and the (undefined) variable b.

```
In[10]:= a = 17
```

```
Out[10]= 17
```

```
In[11]:= -b (a^3 + a - 15)
```

```
Out[11]= -4915 b
```

To work with the output of the previous command, use the special variable % (percentage sign).

Example 12.6. We compute the square of the output of the previous example.

```
In[12]:= %^2
```

```
              2
Out[12]= 24157225 b
```

Mathematica also performs matrix calculations. Matrices are entered with braces and are stored as lists of lists.

Example 12.7. We define two matrices:

$$M = \begin{bmatrix} 1 & 2 & 3 \\ 4 & 5 & 6 \\ 7 & 8 & 9 \end{bmatrix} \text{ and } N = \begin{bmatrix} 0 & 1 & 0 \\ 0 & 0 & 1 \\ 1 & 0 & 0 \end{bmatrix}.$$

```
In[13]:= m = {{1, 2, 3}, {4, 5, 6}, {7, 8, 9}};
```

```
In[14]:= n = {{0, 1, 0}, {0, 0, 1}, {1, 0, 0}};
```

Note. The ; (semicolon) symbol is used to separate commands, allowing you to perform more than one calculation on a line. If you end a command with a semicolon, the output will not be displayed.

Example 12.8. We add and multiply the matrices.

```
In[15]:= m + n
```

```
Out[15]= {{1, 3, 3}, {4, 5, 7}, {8, 8, 9}}
```

```
In[16]:= m . n
```

```
Out[16]= {{3, 1, 2}, {6, 4, 5}, {9, 7, 8}}
```

We may want to use the same variables (e.g., a, m, and n in the above computations) later in a different context. Therefore, it is a good idea to "clear" the values of variables when we are finished using them. We can then check that the values of these variables have disappeared.

```
In[17]:= Clear[a,m,n]

In[18]:= {a, m, n}

Out[18]= {a, m, n}
```

Maple as a calculator

Maple syntax is different from Mathematica syntax, but some things are only slightly different. For instance, to form the product of two expressions, we must explicitly use an asterisk (∗) in Maple. Other things are exactly opposite. For instance, in Maple, ENTER evaluates an expression, whereas SHIFT+ENTER merely breaks the line (i.e., inserts a soft break).

Historically, expressions in Maple had to be "terminated" with a semicolon (;). Newer versions no longer have this requirement, although the semicolon is still allowed. We have kept with the old convention to stay compatible with all versions of Maple.

Example 12.9. We add 2 and 2 to get ... 4.

```
> 2 + 2;
```
$$4$$

To multiply two numbers, type the numbers with an asterisk (∗) between them. Use a caret (^) for exponentiation. Notice that Maple can handle very large numbers easily, even numbers with hundreds of digits.

Example 12.10. A product and the value of 3^{100}.

```
> 1024*59049;
```
$$60466176$$
```
> 3^100;
```
$$515377520732011331036461129765621272702107522001$$

The values of important mathematical constants (such as π and i) are stored in Maple. They are capitalized to distinguish them from variables you might create yourself (e.g., `Pi`, and `I`). Notably, e is *not* represented by `E` in Maple; you must use `exp(1)` instead.

The built-in constants are handled algebraically, but we can request the numerical value of expressions with the `evalf()` function, which evaluates them as "floating point" values.

Example 12.11. Calculations with π, e, and i.

```
> Pi;
```
$$Pi$$
```
> evalf( Pi );
```
$$3.141592654$$
```
> exp( 1 );
```
$$e$$
```
> I*I;
```
$$-1$$

If you want a numerical result given to a high degree of accuracy, include a second argument in the `evalf()` function. The first argument is the number to be calculated. The second argument is the number of decimal places to which the number is computed.

Example 12.12. We calculate π to 100 decimal places.

```
> evalf( Pi, 100 );
3.1415926535897932384626433832795028841971693993751058 \
    2097494459230781640628620899862803482534211708
```

Note. You can obtain information about a specific command by typing a question mark (?) followed by the name of the command. For instance, to find out about the function evalf(), type ? evalf.

In addition to processing numerical calculations, Maple performs algebraic operations. If a variable has not been assigned a value, Maple will work with it algebraically.

Example 12.13. We set a equal to 17, and then calculate with a and the (undefined) variable b.

```
> a := 17;
```
$$a:=17$$
```
> -b*(a^3 + a - 15);
```
$$-4915\ b$$

If you want to work with the output of the previous command, use the special variable % (percentage sign).

Example 12.14. We compute the square of the output of the previous example.

```
> %^2;
```
$$24157225\ b^2$$

Maple can also perform matrix calculations. Many of Maple's matrix functions are part of a package called linalg, which must be loaded first.

Example 12.15. Loading Maple's linear algebra package.

```
> with( linalg ):
```

Note. Lines in Maple may be terminated with a colon (:) to suppress the output of the calculation.

Matrices are entered as lists of lists (in square brackets) using the matrix() function. Matrix operations must occur inside of evalm if we expect them to be evaluated in the usual way. Notice that Matrix multiplication is somewhat special; it uses the &* operator and not the usual *.

Example 12.16. Matrices in Maple.

```
> m := matrix( [ [ 1, 2, 3], [4, 5, 6], [7, 8, 9 ] ] ):
> n := matrix( [ [ 0, 1, 0], [ 0, 0, 1], [ 1, 0, 0 ] ] ):
> m;
```
$$m$$
```
> evalm( m );
```
$$\begin{bmatrix} 1 & 2 & 3 \\ 4 & 5 & 6 \\ 7 & 8 & 9 \end{bmatrix}$$
```
> evalm( m+n );
```
$$\begin{bmatrix} 1 & 3 & 3 \\ 4 & 5 & 7 \\ 8 & 8 & 9 \end{bmatrix}$$

```
> evalm( m &* n );
```

$$\begin{bmatrix} 3 & 1 & 2 \\ 6 & 4 & 5 \\ 9 & 7 & 8 \end{bmatrix}$$

We may want to use the same variables (e.g., a) later in a different context. Therefore, it's a good idea to "unassign" the values of variables when we're finished using them. We can then check that the values of these variables have disappeared.

```
> a := 17;
```
                                        a:=17
```
> unassign( 'a' );
> a;
```
                                          a

Maxima as a calculator

Maxima expects all expressions to end with a semicolon (;), and if you use "command line Maxima," it will continue to prompt for input (without evaluating) until you enter a semicolon. In this respect, Maxima is not unlike many computer languages such as C or Java. This behavior can be convenient as well, since long computations may be split over several lines. The graphical front end, wxMaxima, uses SHIFT+ENTER to run a calculation, like Mathematica does, and will supply a terminating semicolon if you forget one.

Example 12.17. Here we do some simple addition and subtraction. The subtraction is split over two lines.

```
(%i1) 2+2;
(%o1)                           4
(%i2) 3 -
1 ;
(%o2)                           2
```

As you use Maxima, you will notice that the input prompt increments each time you enter a calculation: first (%i1), then (%i2). Each piece of output is also marked, beginning with (%o1). As long as the program remains running, these names can be used as variables in later computations. In a similar spirit, the percent sign alone (%) always refers to "the last answer."

Example 12.18. Compute with the last answer, or with an arbitrary previous answer.

```
(%i1) 2+2;
(%o1)                           4
(%i2) 3 * %;
(%o2)                          12
(%i3) 5 * %i1;
(%o3)                          20
```

Maxima can handle very large numbers, larger than may be convenient to work with on a hand-held calculator.

Example 12.19. We evaluate 3^{100}.

```
(%i4) 3^100;
(%o4)              515377520732011331036461129765621272702107522001
```

Many standard constants, such as π, e, and the imaginary unit i, are part of Maxima. Maxima uses a syntax with percent signs to signify the built-in constants, very much like the special variables that represent the results of previous calculations. So %pi, %e, and %i denote π, e, and i.

Maxima handles constants algebraically when possible, but we can force numerical presentation with the special symbol numer.

Example 12.20. Calculations with π, e, and i.

```
(%i1) %pi;
(%o1)                                %pi
(%i2) %pi,numer;
(%o2)                           3.141592653589793
(%i3) %e,numer;
(%o3)                           2.718281828459045
(%i4) %i * %i;
(%o4)                                - 1
```

To specify arbitrary "big floating point" precision, there is also a special symbol called bfloat.

Example 12.21. The number π calculated to 100 decimal places.

```
(%i1) %pi,bfloat,fpprec=100;
(%o1) 3.141592653589793238462643383279502884197169399375\
51058209749445923078164062862089986280348253421170680b0
```

Note. Big floating point results are always presented in scientific notation, so do not forget to read the exponent following the "b". The result above is in fact $3.14\ldots \times 10^0$.

Note. You can obtain information about a specific symbol (or function) by typing a question mark (and space) followed by the name of the symbol. This is one of the few times Maxima does *not* want a terminating semicolon. For example, to find out about the special symbol numer:

```
(%i1) ? numer

 -- Special symbol: numer
     'numer' causes some mathematical functions (including
     exponentiation) with numerical arguments to be evaluated
     in floating point. It causes variables in 'expr' which
     have been given numerals to be replaced by their values.
     It also sets the 'float' switch on.

 There are also some inexact matches for 'numer'.
 Try '?? numer' to see them.

(%o1)                                true
```

In addition to numerical calculations, Maxima can work with variables and do algebraic operations. The colon (:) is the assignment operator. We use it to define variables. If a variable has not been assigned a value, Maxima will work with it algebraically.

Example 12.22. We set a equal to 17, and then calculate with a and the (undefined) variable b.

```
(%i1) a : 17;
(%o1)                              17
(%i2) -b * (a^3 + a - 15);
(%o2)                          - 4915 b
```

We can tell Maxima that we wish it to handle a variable name algebraically, even if it knows how to evaluate it, by using the single quote (') symbol. For example, in an expression, 'x will not be expanded, even if a value of x has already been defined. When we are ready to have Maxima evaluate an expression, we can use the ev() function.

Example 12.23. Quoting a variable prevents evaluation.

```
(%i1) x : 2;
(%o1)                               2
(%i2) y : x^2;
(%o2)                               4
(%i3) z : 'x^2;
                                    2
(%o3)                               x
(%i4) ev(z);
(%o4)                               4
```

Maxima can handle vectors and matrices that contain values or variables, and it can do the usual mathematical operations on them. Simple row vectors may be entered as lists in square brackets. Matrices with more than one row are defined using the matrix() function, which takes the rows of the matrix as its arguments.

Note. The asterisk (*) and caret (^) operators both work component-wise on matrices (i.e., on each entry). Use dot (.) and double caret (^^) for matrix multiplication and matrix exponentiation, respectively.

Example 12.24. Matrices in Maxima.

```
(%i1) M : matrix( [ 17, b], [1, 17 ] );
                              [ 17  b  ]
(%o1)                         [        ]
                              [ 1    17 ]
(%i2) M . M;
                        [ b + 289    34 b   ]
(%o2)                   [                   ]
                        [   34     b + 289 ]
(%i3) M + M;
                              [ 34  2 b ]
(%o3)                         [         ]
                              [ 2    34 ]
(%i4) [ 1, 0] . M;
(%o4)                          [ 17  b ]
```

(%i5) M ^^ 2;

```
                          [ b + 289     34 b   ]
(%o5)                     [                     ]
                          [   34      b + 289 ]
```

The kill() function is used to clear variables. For example, if we wish to clear the variable M to use it in some other way, we can kill it.

Example 12.25. Clearing (killing) the variable M.

```
(%i6) kill( M );
(%o6)                                   done
(%i7) M;
(%o7)                                    M
```

12.3　How to compute functions

Functions in Mathematica

The operator N, which we saw earlier, is actually a function. Mathematica contains many such built-in functions, and you can usually guess the names of common functions. For instance, Sin[x] computes $\sin x$.

Note. In Mathematica, every built-in function name begins with a capital letter. Arguments of functions follow in square brackets.

Example 12.26. We calculate $\sin(\pi/2)$ and the binomial coefficient $\binom{7}{2}$.

```
In[1]:= Sin[Pi/2]

Out[1]= 1

In[2]:= Binomial[7,2]

Out[2]= 21
```

Some functions have outputs that are lists.

Example 12.27. The command FactorInteger determines the prime factorization of an integer. Here we find the prime factorization of the number 60466176.

```
In[3]:= FactorInteger[60466176]

Out[3]= {{2, 10}, {3, 10}}
```

The output tells us that $60466176 = 2^{10} \cdot 3^{10}$.

In the next example, we calculate and display a table of function values.

Example 12.28. The function Prime[n] gives the nth prime number. Using this function, we construct a table of the first 100 prime numbers.

Command	Meaning	Example Input	Meaning
Sqrt[]	square root	Sqrt[5]	$\sqrt{5}$
Exp[]	exponential	Exp[x]	e^x
Log[]	natural logarithm	Log[10]	$\ln 10$
Log[,]	logarithm	Log[10,5]	$\log_{10} 5$
Sin[]	sine	Sin[x]	$\sin x$
Cos[]	cosine	Cos[x]	$\cos x$
Tan[]	tangent	Tan[x]	$\tan x$
Sum[,]	sum	Sum[a[i],{i,1,n}]	$\sum_{i=1}^{n} a_i$
Product[,]	product	Product[a[i],{i,1,5}]	$\prod_{i=1}^{5} a_i$
Mod[,]	modulus	Mod[10,3]	$10 \bmod 3$

TABLE 12.1: Some Mathematica functions.

```
In[4]:= Table[Prime[n], {n, 1, 100}]

Out[4]= {2, 3, 5, 7, 11, 13, 17, 19, 23, 29, 31, 37, 41, 43,
> 47, 53, 59, 61, 67, 71, 73, 79, 83, 89, 97, 101, 103, 107,
> 109, 113, 127, 131, 137, 139, 149, 151, 157, 163, 167, 173,
> 179, 181, 191, 193, 197, 199, 211, 223, 227, 229, 233, 239,
> 241, 251, 257, 263, 269, 271, 277, 281, 283, 293, 307, 311,
> 313, 317, 331, 337, 347, 349, 353, 359, 367, 373, 379, 383,
> 389, 397, 401, 409, 419, 421, 431, 433, 439, 443, 449, 457,
> 461, 463, 467, 479, 487, 491, 499, 503, 509, 521, 523, 541}
```

Mathematica can evaluate functions both arithmetically and symbolically.

Example 12.29. The sums $\sum_{i=1}^{10} i^2$ and $\sum_{i=1}^{n} i^2$.

```
In[5]:= Sum[i^2, {i, 1, 10}]

Out[5]= 385

In[6]:= Sum[i^2, {i, 1, n}]

        1
Out[6]= - n (1 + n) (1 + 2 n)
        6
```

As we can see, Mathematica knows that $\sum_{i=1}^{n} i^2 = n(n+1)(2n+1)/6$.

Table 12.1 displays some useful Mathematica functions.

You can define your own functions. To create a function $f(x)$, write f[x_] := followed by the definition of f.

Example 12.30. We define a function $f(x) = x^3 + \sin x$.

```
In[7]:= f[x_] := x^3 + Sin[x]
```

```
In[8] := f[Pi/2]
```

$$
\text{Out[8]} = 1 + \frac{Pi^3}{8}
$$

We can differentiate and integrate our function.

```
In[9] := D[f[x],x]
```

$$
\text{Out[9]} = 3\ x^2 + Cos[x]
$$

```
In[10] := Integrate[f[x],x]
```

$$
\text{Out[10]} = \frac{x^4}{4} - Cos[x]
$$

```
In[11] := Integrate[f[x], {x, 0, Pi}]
```

$$
\text{Out[11]} = \frac{1}{4}\ (\ 8 + Pi^4\ )
$$

Note. Mathematica does not supply an additive constant $(+C)$ for indefinite integrals.

You can define functions recursively (in terms of previous values), as with the function below. Notice the use of = for the assignment of initial values in contrast with := for the definition of the iteration.

Example 12.31. We define the Fibonacci sequence.

```
In[1] := f[0] = 1;
```

```
In[2] := f[1] = 1;
```

```
In[3] := f[n_] := f[n] = f[n-2] + f[n-1]
```

```
In[4] := Table[f[n], {n, 0, 10}]
```

```
Out[4] = {1, 1, 2, 3, 5, 8, 13, 21, 34, 55, 89}
```

You may wonder at the construction in line 3. In Mathematica, := is the "delayed" assignment operator as opposed to =, which does immediate assignment. When we use delayed assignment, Mathematica will wait to fully evaluate an expression.

To explore this idea, consider the first time we ask Mathematica for the value f[3]. Since we used :=, the value of 3 will be substituted for n on the right side of the definition, which will evaluate to the expression f[3] = f[1] + f[2] (which is actually an assignment itself). Mathematica already knows the values of f[1] and f[2] and consequently sets f[3] = 2 using the immediate assignment operator. The = also "returns a value," the value that we see.

In more detail, the steps performed by Mathematica to do the evaluation of f[3] are:

1. f[3] = f[3-2] + f[3-1] (this is the right-hand side of the :=, with 3 substituted for n in all places)

2. f[3] = f[1] + f[2]

3. f[3] = 1 + 1 (from previously defined values)

4. f[3] = 2 (this is ready to perform immediate assignment)

5. 2 (the return value of the = assignment).

Of course, all we see is the final result:

```
In[5]:= f[3]
```

```
Out[5] = 2
```

The consequence of doing the computation this way is that Mathematica now knows permanently that f[3] has the value 2 and will never have to evaluate it again (say, when we ask for f[4] or any other value). This becomes important for larger values, like f[100], which would evaluate too slowly if we created the function less carefully.

Functions in Maple

Some of the commands we've seen, like matrix() and evalf(), are actually functions. Maple contains many built-in functions, including common mathematical functions like sin() and cos().

Example 12.32. We calculate $\sin(\pi/2)$ and the binomial coefficient $\binom{7}{2}$.

```
> sin( Pi/ 2 );
                                1
> binomial( 7, 2 );
                               21
```

Maple also has functions related to Number Theory. For example, ifactor() determines the prime factorization of an integer.

```
> ifactor( 60466176 );
                            10      10
                          (2)    (3)
```

Maple can easily generate sequences (i.e., a table or list of values). Using the seq() function and the ithprime() function (which returns the ith prime number), we construct a list of the first 100 prime numbers.

```
> seq( ithprime(n), n=1..100 );
2, 3, 5, 7, 11, 13, 17, 19, 23, 29, 31, 37, 41, 43, 47, 53, 59, 61,
   67, 71, 73, 79, 83, 89, 97, 101, 103, 107, 109, 113, 127, 131, 137,
   139, 149, 151, 157, 163, 167, 173, 179, 181, 191, 193, 197, 199,
```

Command	Meaning	Example Input	Meaning
`sqrt()`	square root	`sqrt(5)`	$\sqrt{5}$
`exp()`	exponential	`exp(x)`	e^x
`ln()` or `log()`	natural logarithm	`ln(10)` or `log(10)`	$\ln 10$
`log10()`	common logarithm	`log10(5)`	$\log_{10} 5$
`sin()`	sine	`sin(x)`	$\sin x$
`cos()`	cosine	`cos(x)`	$\cos x$
`tan()`	tangent	`tan(x)`	$\tan x$
`sum(,)`	sum	`sum(i^2,i=1..n)`	$\sum_{i=1}^{n} i^2$
`mul(,)`	product	`mul(i^2,i=1..5)`	$\prod_{i=1}^{5} i^2$
`mod(,)`	modulus	`mod(10,3)`	$10 \bmod 3$

TABLE 12.2: Some Maple functions.

211, 223, 227, 229, 233, 239, 241, 251, 257, 263, 269, 271, 277, 281, 283, 293, 307, 311, 313, 317, 331, 337, 347, 349, 353, 359, 367, 373, 379, 383, 389, 397, 401, 409, 419, 421, 431, 433, 439, 443, 449, 457, 461, 463, 467, 479, 487, 491, 499, 503, 509, 521, 523, 541

Maple has functions that compute sums and products, which it can evaluate both arithmetically and symbolically.

Example 12.33. The sums $\sum_{i=1}^{10} i^2$ and $\sum_{i=1}^{n} i^2$.

```
> sum( i^2, i=1..10 );
```
$$385$$
```
> sum( i^2, i=1..n );
```

```
            3          2
       (n+1)      (n+1)     n    1
       -----   -  -----  +  -  + -
         3          2       6    6
```

Maple knows in its own way that $\sum_{i=1}^{n} i^2 = n(n+1)(2n+1)/6$.

Table 12.2 displays some useful Maple functions.

You can create your own functions. To create a function, use the `:=` and `->` operators.

Example 12.34. We define a function $f(x) = x^3 + \sin x$.

```
> f := x -> x^3 + sin(x);
```

```
                         3
           f := x -> x  + sin(x)
```
```
> f( Pi/2 );
```

```
              3
            Pi
            --- + 1
             8
```

We can differentiate and integrate our function.

```
> diff( f(x), x );
                                  2
                             3 x  + cos(x)
> integrate( f(x), x );
                                4
                               x
                               --- - cos(x)
                                4
> integrate( f(x), x=0..Pi );
                                  4
                                Pi
                                --- + 2
                                 4
```

Note. Maple does not supply an additive constant ($+C$) for indefinite integrals.

Functions in Maxima

Some of the commands we've seen, like `matrix()` and `kill()`, are actually functions. Maxima contains many built-in functions, including common mathematical functions like `sin()` and `cos()`.

Example 12.35. We compute a couple of trigonometric functions.

```
(%i1) sin( %pi/2 );
(%o1)                             1
(%i2) cos( %pi/4 );

                                  1
(%o2)                          -------
                               sqrt(2)
```

Maxima also contains functions for counting things, like binomial coefficients, as well as for dealing with numbers and factorizations.

Example 12.36. Computing with some of Maxima's other functions:

```
(%i1) binomial( 7, 3 );
(%o1)                            35
(%i2) factor( 210 );
(%o2)                          2 3 5 7
(%i3) next_prime(1);
(%o3)                             2
(%i4) next_prime(8);
(%o4)                            11
(%i5) prev_prime(2009);
(%o5)                           2003
```

Some functions must be loaded before they are available. For example, by default Maxima does not load the `permutation()` function. It is contained in a packaged called `functs`. Similarly, it does not load all of the functions you might use to do descriptive statistics (like means or standard deviations).

Example 12.37. Loading functions into Maxima:

```
(%i8) load( functs );
(%o8)       /usr/share/maxima/5.13.0/share/simplification/functs.mac
(%i9) permutation( 10, 3 );
(%o9)                             720
(%i10) load (descriptive)$
(%i11) mean( [ 1, 2, 3, 4, 5 ] );
(%o11)                             3
```

Note. The dollar sign ($) at the end of a line serves exactly as a semicolon, but will suppress the output. Notice that there is no (%o10) above.

Maple has functions that compute sums and products, and it can evaluate them both arithmetically and symbolically.

Example 12.38. Maxima can compute products and sums:

```
(%i17) prod( sqrt(i), i, 1, 4 );
                              3/2
(%o17)                       2      sqrt(3)
(%i18) sum( i^2, i, 1, n );

                              n
                            ====
                            \       2
(%o18)                       >     i
                            /
                            ====
                            i = 1
```

Note. The variable `simpsum` controls whether Maxima will perform simplifications to sums. It is false by default. Setting it to true will make Maxima return a closed form for the sum above.

Example 12.39. Maxima will simplify sums when `simpsum` is true:

```
(%i19) simpsum : true;
(%o19)                            true
(%i20) sum( i^2, i, 1, n );
                              3       2
                           2 n  + 3 n  + n
(%o20)                     ---------------
                                  6
```

Maxima knows that $\sum_{i=1}^{n} i^2 = n(n+1)(2n+1)/6$.

Maxima can also apply a condition to a calculation. For example, if we are interested in the previous sum when $n = 10$, we could reevaluate it with that condition placed after the comma (,) operator.

Command	Meaning	Example Input	Meaning
`sqrt()`	square root	`sqrt(5)`	$\sqrt{5}$
`exp()`	exponential	`exp(x)`	e^x
`log()`	natural logarithm	`log(10)`	$\ln 10$
`sin()`	sine	`sin(x)`	$\sin x$
`cos()`	cosine	`cos(x)`	$\cos x$
`tan()`	tangent	`tan(x)`	$\tan x$
`sum(,,,)`	sum	`sum(i^2,i,1,n)`	$\sum_{i=1}^{n} i^2$
`prod(,,,)`	product	`prod(i^2,i,1,5)`	$\prod_{i=1}^{5} i^2$
`mod(,)`	modulus	`mod(10,3)`	$10 \bmod 3$

TABLE 12.3: Some Maxima functions.

Example 12.40. Evaluating an expression at a particular value:

```
(%i21) %, n=10;
(%o21)                          385
```

Table 12.3 displays some useful Maxima functions.

To define your own functions, use the `:=` operator.

Example 12.41. We define a function $f(x) = x^3 + \sin x$.

```
(%i1) f(x) := x^3 + sin(x);
                                3
(%o1)                   f(x) := x  + sin(x)
(%i2) f( %pi/2 );
                              3
                           %pi
(%o2)                      ---- + 1
                            8
```

We can differentiate and integrate our function.

```
(%i3) diff( f(x), x );
                                      2
(%o3)                     cos(x) + 3 x
(%i4) integrate( f(x), x );
                            4
                           x
(%o4)                      -- - cos(x)
                           4
(%i5) integrate( f(x), x, 0, %pi );
                              4
                           %pi  + 8
(%o5)                      --------
                              4
```

Note. Maxima does not supply an additive constant $(+C)$ for indefinite integrals.

Functions in Maxima may also be defined recursively (in terms of previous values). Perhaps the easiest way is to code the function as a small program. For example, the famous Fibonacci sequence might be defined like this:

Example 12.42. Defining and evaluating the Fibonacci sequence.

```
(%i1) f(n) := if n<2 then 1 else f(n-1)+f(n-2);
(%o1)              f(n) := if n < 2 then 1 else f(n - 1) + f(n - 2)
(%i2) makelist( f(n), n, 1, 10 );
(%o2)                    [1, 2, 3, 5, 8, 13, 21, 34, 55, 89]
```

Although technically correct, this is a very inefficient way to compute Fibonacci numbers. The problem is that the same values of the Fibonacci sequence must be computed over and over again (the evaluation of f(3) computes f(1) twice, for example). Large values like $f(100)$ will take much too long to compute using this naive definition. From a performance point of view, it would have been better to use a loop (or some even better method) instead of recursion for this particular computation.

12.4   How to make graphs

Graphs in Mathematica

Mathematica offers many graphing options. We show a few examples here. You can create graphs of functions using Mathematica's `Plot` command.

Example 12.43. A graph of the function $y = \sin x$, for $0 \le x \le 2\pi$.

```
In[1]:= Plot[Sin[x], {x, 0, 2 Pi}]
```

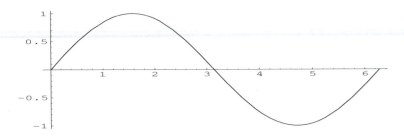

You can graph several curves together.

Example 12.44. A graph of three lines, $y = 4x + 1$, $y = -x + 4$, and $y = 9x - 8$, for $0 \le x \le 2$.

```
In[1]:= f[x_] := 4 x + 1;
```

```
In[2]:= g[x_] := -x + 4;
```

```
In[3]:= h[x_] := 9 x - 8;
```

```
In[4]:= Plot[{f[x], g[x], h[x]}, {x, 0, 2}]
```

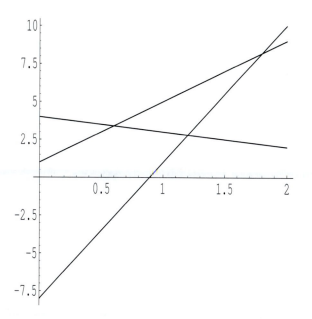

You can create 3-dimensional graphs of surfaces using the `Plot3D` command.

Example 12.45. A graph of the surface $z = e^{-(x^2+y^2)}$, for $-2 \leq x, y \leq 2$.

```
In[1]:= Plot3D[E^(-(x^2 + y^2)), {x, -2, 2}, {y, -2, 2}]
```

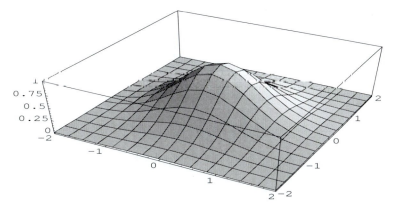

Example 12.46. We plot a sphere using parametric equations.

```
In[1]:= ParametricPlot3D[{Sin[phi] Cos[theta],
 Sin[phi] Sin[theta], Cos[phi]},
 {phi, 0, Pi}, {theta, 0, 2 Pi}]
```

See Figure 2 in the color insert.

If you want an Encapsulated PostScript (EPS) version of your image, use an `Export` command.

```
In[2]:= Export["newgraph.eps", %]
```

The graphics file, `newgraph.eps`, is stored in a "working directory," which you can identify using the `Directory` command.

```
In[3]:= Directory[]

Out[3]= C:\\Program Files\\Wolfram Research\\Mathematica\\7
```

Graphs in Maple

Maple can create beautiful graphs, and a few examples are shown here. You can create graphs of functions using the `plot()` function.

Example 12.47. A graph of the function $y = \sin x$, for $0 \le x \le 2\pi$.

```
> plot( sin(x), x=0..2*Pi );
```

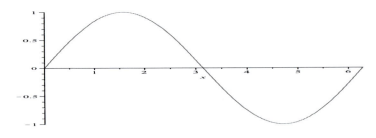

You can graph several curves together by passing a set of functions to the `plot()` function.

Example 12.48. A graph of three lines, $y = 4x + 1$, $y = -x + 4$, and $y = 9x - 8$, for $0 \le x \le 2$.

```
> f := x -> 4*x+1:
> g := x -> -x + 4:
> h := x -> 9*x - 8:
> plot( { f(x), g(x), h(x) }, x=0..2 );
```

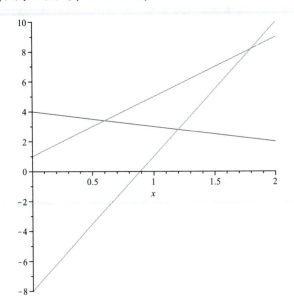

You can create 3-dimensional graphs of surfaces using the `plot3d()` function.

Example 12.49. A graph of the surface $z = e^{-(x^2+y^2)}$, for $-2 \leq x, y \leq 2$.

```
> plot3d( exp( -(x^2 + y^2)), x=-2..2, y=-2..2 );
```

See Figure 3 in the color insert.

The `plot3d()` function can also perform parametric plots if you pass a list (in square brackets) of the (x, y, z) coordinates of your surface.

Example 12.50. A parametric plot of a torus.

```
> plot3d([(2+cos(v))*cos(u), (2+cos(v))*sin(u), sin(v)],
u = 0 .. 2*Pi, v = 0 .. 2*Pi,
axes = framed, labels = [x, y, z], scaling = constrained);
```

See Figure 4 in the color insert.

Images generated in Maple may be saved in several different formats. Simply right-click the image you wish to save and choose Export.

Graphs in Maxima

Maxima can graph both curves and surfaces. You can create graphs of curves with the `plot2d()` command.

Example 12.51. A graph of the function $y = \sin x$, for $0 \leq x \leq 2\pi$.

```
(%i1) plot2d( sin(x), [x, 0, 2*%pi ] )$
```

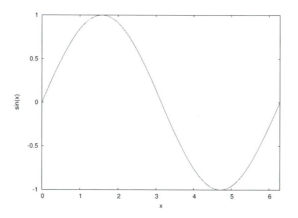

Note. If you use wxMaxima, substitute `wxplot2d()` for `plot2d()` to create "inline" graphics that are integrated into your document.

You can graph several curves together by combining them in a list (i.e., in square brackets).

Example 12.52. A graph of three lines, $y = 4x + 1$, $y = -x + 4$, and $y = 9x - 8$, for $0 \le x \le 2$

```
(%i1) f(x) := 4*x + 1;
(%o1)                              f(x) := 4 x + 1
(%i2) g(x) := -x + 4;
(%o2)                              g(x) := - x + 4
(%i3) h(x) := 9*x - 8;
(%o3)                              h(x) := 9 x - 8
(%i4) plot2d( [f(x), g(x), h(x)], [x,0,2] )$
```

See Figure 5 in the color insert.

You can create 3-dimensional graphs of surfaces using the `plot3d()` command.

Example 12.53. A graph of the surface $z = e^{-(x^2+y^2)}$, for $-2 \le x, y \le 2$.

```
(%i1) plot3d( %e^( -(x^2+y^2) ), [x,-2,2], [y,-2,2] )$
```

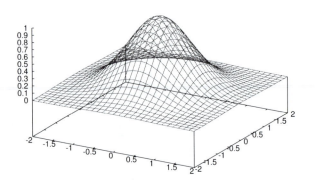

Note. In wxMaxima documents, use `wxplot3d()` to create inline 3D plots.

The plotting engine that lies beneath Maxima is called gnuplot, and it is a very functional program of its own. By adding extra arguments to the `plot3d()` command, we can ask gnuplot to enhance the graphs that we construct in Maxima. For example, we can choose different coloring algorithms, we can draw contour lines on the surface (or beneath it), we can specify a particular viewing angle, or we can label the axes.

Example 12.54. A nicely colored surface with contour lines and labeled axes.

```
(%i1) plot3d( %e^( -(x^2+y^2) ), [x,-2,2], [y,-2,2],
[gnuplot_preamble, "set pm3d; set hidden3d; set contour both;\
 set xlabel 'x'; set ylabel 'y'; set zlabel 'z';\
 set key top left"] )$
```

See Figure 6 in the color insert.

Gnuplot settings that you may wish to include:

• set contour base

- `set contour surface`

- `set contour both`

- `set cntrparam levels 10`

- `set cntrparam levels discrete 1,2,4,8` (sets contour lines at specific values)

- `set hidden3d` (makes the surface opaque)

- `set nohidden3d` (makes the surface transparent)

- `set key` (moves the legend key to different places)

- `set pm3d` (colors the surface according to height)

- `set surface` (draws the surface)

- `set nosurface` (hides the surface, but contour lines and pm3d will still show)

- `set xlabel 'x axis'`

- `set ylabel 'y axis'`

- `set zlabel 'z axis'`

- `set view 45,115` (rotates the viewpoint around x-axis and z-axis)

Maxima doesn't have a command for printing a plot, but it does have options for saving a plot to a file, which you can then print. The most useful file formats are probably PostScript and PNG (Portable Network Graphics). Since gnuplot actually does the plotting, we simply have Maxima tell gnuplot what we want. If we don't specify, plots will be saved with the name `maxplot.ps` (for PostScript files) or `maxplot.png` (for PNG files).

Example 12.55. Saving a plot to a file.

```
(%i1) plot2d( x^2, [x,0,3], [gnuplot_term, ps] )$
```

```
(%i2) plot2d( x^2, [x,0,3], [gnuplot_term, png] )$
```

Maxima will try to guess the best place to save your files, usually your "home" directory on GNU/Linux or Unix systems and in the "My Documunts" folder on Windows systems. The variable `maxima_tempdir` determines where images will be saved. Notice that even on systems running Microsoft Windows, Maxima uses forward slashes to indicate folders (just as browsers do with Web addresses).

Example 12.56. Changing the default location for saved images.

```
(%i3) maxima_tempdir;
(%o3)                     C:/Documents and Settings/dbindner
(%i4) maxima_tempdir : "C:/";
(%o4)                           C:/
```

We can also specify a file name if we wish. For example, if we want a plot of $y = x^2$ saved as (encapsulated) PostScript with the file name `parabola.eps`, we could do something like this:

Example 12.57. Naming the file that a Maxima plot is saved to.

```
(%i5) plot2d( x^2, [x,0,3],
      [gnuplot_term, ps], [gnuplot_out_file, "parabola.eps" ] )$
```

12.5 How to do simple programming

Programming in Mathematica

Mathematica supports a full spectrum of programming paradigms, including procedural, functional, transformational, and object-oriented approaches. We give a sampling here.

A Do loop is a simple kind of program that performs a calculation some fixed number of times.

Example 12.58. A calculation related to the fractal known as the Mandelbrot set. We set $c = -0.5 + 0.5i$ and $z = 0 + 0i$. Then we iterate the function $f(z) = z^2 + c$ ten times.

```
In[1]:= c = -0.5 + 0.5 I;

In[2]:= z = 0 + 0 I;

In[3]:= Do[z = z^2 + c, {10}];

In[4]:= z

Out[4]= -0.11932 + 0.219608 I

In[5]:= Clear[c,z]
```

This is fine for a one-time computation. But perhaps we wish to run the same program several times, with different values for c and different numbers of iterations. To do this, we create a module, which is a procedure containing local variables.

Example 12.59. We define a module containing the local variable z. The values of c and i (the number of iterations) are input when the module is called.

```
In[1]:= f[c_, i_] := Module[{z}, z = 0 + 0 I;
            Do[z = z^2 + c, {i}];
            z
         ]

In[2]:= f[-0.5 + 0.5 I, 10]

Out[2]= -0.11932 + 0.219608 I
```

Notice that z has no value outside the module.

```
In[3]:= z

Out[3]= z
```

It is good programming practice to use modules, and to make them small and easy to understand.

One useful feature of functions in Mathematica is that they are "threaded" over lists automatically and applied to each list item.

Example 12.60. We thread addition and cubing operations over the list $\{a, b, c\}$.

```
In[1]:= a := 6; c := 2+I;

In[2]:= 1000 + {a,b,c}^3
```

```
                3
Out[2]= {1216, 1000 + b , 1002 + 11 I}
```

Sometimes functions are complex enough to be called programs.

Example 12.61. We define a function

$$f(n) = \frac{1}{n} \sum_{k|n} \phi(k) 2^{n/k}.$$

Note. From the Pólya theory of counting, $f(n)$ is the number of distinct (up to rotation and flipping) necklaces formed by n beads of two types.

The summation is over a set of numbers, namely, the set of positive divisors of n. This set is obtained in Mathematica as `Divisors[n]`. We need to apply the summand, $\phi(k)2^{n/k}$, to each element of this set. The summand contains a "dummy variable," k. To define the summand as a Mathematica function, we replace each instance of the dummy variable with the marker # (number sign).

```
EulerPhi[#]2^(n/#)&
```

The & (ampersand) identifies the function as a "pure function" in which the argument is denoted by #.

Then we apply the function to the set `Divisors[n]` as follows.

```
EulerPhi[#]2^(n/#)&/@Divisors[n]
```

(The construction `f/@s` applies a function `f` to a set `s`.)

Finally, we add the elements of the set produced by this process. The expression `Plus@@s` adds the elements of the set `s`. Thus, we can now define our function in Mathematica.

```
In[1]:= f[n_] := (1/n)Plus@@(EulerPhi[#]2^(n/#)&/@Divisors[n])
```

We test our function.

```
In[2]:= f[4]
```

```
Out[2]= 6
```

It is easy to verify by inspection that there are exactly six different necklaces made of four beads of two types.

And now we compute a large value of the function.

```
In[3]:= f[100]
```

```
Out[3]= 1267650600228230527396813560
```

Programming in Maple

Programming in Maple is similar to programming in Mathematica. In Maple, programs are called procedures. Let's create a procedure to find the greatest common divisor (gcd) of two positive integers. Recall that the Euclidean algorithm for finding the gcd of integers a and b is based on the relation

$$\gcd(a, b) = \gcd(b, r),$$

where $a = bq + r$, with $0 \leq r < b$. In Maple, the remainder r is given by `modp(a,b)`.

Here is our procedure, which we call **ourgcd**.

```
> ourgcd := proc(a,b)
  local atemp, btemp;
  (atemp, btemp) := (a, b);
  while btemp > 0 do
    (atemp, btemp) := (btemp,modp(atemp,btemp));
  end do;
  atemp;
end proc:
```

We use the construction `proc()` and `end proc` to begin and end a procedure. The inputs in the procedure are **a** and **b**. In the second line, we declare the variables to be used in the procedure (**atemp** and **btemp**). In the third line, we set values for these variables. The fourth line begins with a `while` loop, which has a test condition (**btemp > 0**) and a command to be performed (the key step). The command is bracketed by `do` and `end do`. Finally, an output (the value of **atemp**) is output.

```
> ourgcd(15,24);
```
$$3$$

Programming in Maxima

Interesting and powerful programs may be expressed in Maxima. In fact, Maxima can be programmed in two different ways. Not only is Maxima a full-fledged programming language, but Maxima itself is written in Lisp, and it is possible to graft custom Lisp programs into Maxima. Both simple and elaborate programs can generally be written without resorting to Lisp, however, building on the syntax that we have already learned.

A `thru-do` loop is a simple kind of program that performs a calculation some fixed number of times.

Example 12.62. A calculation related to the fractal known as the Mandelbrot set. We set $c = -0.5 + 0.5i$ and $z = 0$. Then we iterate $f(z) = z^2 + c$ ten times.

```
(%i1) c : -0.5 + 0.5*%i;
(%o1)                          0.5 %i - 0.5
(%i2) z : 0;
(%o2)                               0
(%i3) thru 10 do z : z^2 + c;
(%o3)                             done
(%i4) expand(z);
(%o4)          0.21960760831735 %i - 0.11932015744635
```

This is fine for a one-time computation. But perhaps we wish to run the same program several times, using different values of c and for a different number of iterations. To do this, we can create a function to perform the computation.

An essential part of getting complex functions to work well is the `block()` function, which is used in Maxima to encapsulate a program. The `block()` function does two important tasks that help protect our code.

The first useful thing that `block()` does is provide a new variable name space. If the first argument in a block is a list of variable names (or variable initializations), then those variables will be local to the block. A local variable z, for example, used in a program will not overwrite a variable z you may have been using in a calculation elsewhere.

The second task that `block()` does is to evaluate a compound expression and return the value of the last computation performed. Whatever value is calculated by the last step of a program becomes the return value for the function.

Example 12.63. A function with one local variable, z, encapsulated in a block. Notice that outside the program z has not been changed or defined.

```
(%i1) f(c,i) := block(
 [ z : 0 ],
 thru i do z : z^2 + c,
 expand(z)
);
(%o1) f(c, i) := block(z : 0, thru i do z : z  + c, expand(z))
(%i2) f( -0.5+0.5*%i, 10 );
(%o2)            0.21960760831735 %i - 0.11932015744635
(%i3) z;
(%o3)                              z
```

Note. The expressions that make up a compound expression are separated by commas, not semicolons, and the last expression in the block is not followed by anything. The only semicolon comes at the end of the block.

Programs in Maxima can be as elaborate as you can imagine and code. Here is a bit longer example. A common task for Calculus students is to learn and compute Newton's method to find zeros of a function. Maxima has native ways to solve for zeros, but for this example we program Newton's method into Maxima directly.

Example 12.64. A hand-coded Newton's method program used to find the square root of 81 (a solution to $x^2 - 81 = 0$).

```
(%i1) newt( f, x, err ) := block(
 [ df, old, i ],          /* local variables */
 df : diff( f('x), 'x ),  /* differentiate algebraically */

 old : x,
 x : ev(x-f(x)/df, numer),/* evaluate numerically */

 for i:1 thru 20 do (
  if abs(x-old) < abs(err) then return(x),

  old : x,
  x : ev(x-f(x)/df, numer)
 )
```

```
/* if loop runs out without a return(), */
/* no value will be returned */
)$
```

```
(%i2) f(x) := x^2 - 81;
```

$$(\%o2) \qquad\qquad f(x) := x^2 - 81$$

```
(%i3) newt( f, 1, .001 );
```

$$(\%o3) \qquad\qquad 9.000000000007093$$

12.6   How to learn more

Learning more about Mathematica

There are many aspects of Mathematica not discussed in this introduction, such as standard and add-on packages, sound, and animated graphics. Here are some resources for you to investigate to learn more.

The definitive book about Mathematica is [59]. Good beginning books are [16] and [56]. For informative examples of Mathematica in a wide variety of settings, see [11], [36], [48], and [58]. The books [27] and [26] show many applications of Mathematica to Calculus. For a comprehensive guide to add-on packages, see [37].

For complete and up-to-date information describing Mathematica, you may want to visit the Web site www.wolfram.com. Another interesting site, concerning the *Mathematica in Education and Research* journal, is www.telospub.com/journal/MIER/.

Learning more about Maple

Maple is produced by Waterloo Maple, Inc. You can obtain more information at www.maplesoft.com.

A favorite source to learn more about Maple is the MaplePrimes discussion site at www.mapleprimes.com.

Learning more about Maxima

The Maxima project page is maxima.sourceforge.net. The project page contains documentation in various languages for Maxima as well as links to sites for downloading Maxima. Most notably, a version of Maxima that works on Microsoft Windows is available as a free download from sourceforge.net/projects/maxima/files/.

Exercises

1. Graph each of the functions. Experiment with different domains or viewpoints to display the best images.

(a) $f(x) = \dfrac{x}{1+x^2}$

(b) $y = x\sin(1/x)$

(c) $g(x,y) = \cos(x) + \sin(y)$

(d) $z = \dfrac{xy}{x^2+y^2}$

2. Let $f(x) = \dfrac{x}{1+x^2}$.

   (a) Find $f'(x)$ and $f''(x)$.

   (b) Find $f'(-1)$ and $f'(0)$.

   (c) Find $f''(0)$ and $f''(1)$.

3. Find the prime factorization of each integer.

   (a) 3,527,218,133,309,949,276,293

   (b) 471,945,325,930,166,269

   (c) 471,945,325,930,166,281

4. Compute each expression. Do you notice a pattern?

   (a) $3^6 \bmod 7$

   (b) $6^{10} \bmod 11$

   (c) $7^{20} \bmod 21$

   (d) $7^{22} \bmod 23$

5. In 1976, Whitfield Diffie and Martin Hellman published a way for two people to share a secret number when communicating over an insecure medium (like the Internet).

If Alice and Bob want to communicate, Alice picks a random prime p and another random number g smaller than p. To create a shared secret, Alice and Bob each pick part of the secret. Alice picks a random number a and Bob picks a random b. Alice computes $A = g^a \bmod p$, which she sends to Bob, and Bob computes $B = g^b \bmod p$ to send to Alice. The actual secret is $g^{ab} \bmod p$, which both Alice and Bob can compute, but which someone watching only the communication would find difficult to reproduce (because it's difficult to figure out what a and b are).

   (a) Play the part of Alice. We'll use $p = 36479$ for our prime and a random number $g = 14$, which she shares with Bob (and potentially with the world). Then Alice needs a secret value, and we'll use $a = 5013$. Alice sends Bob the value $A = g^a \bmod p$. Compute $14^{5013} \bmod 36479$ to send to Bob.

   (b) Play the part of Bob, who has received p, g, A from Alice. Bob needs a secret value, and we'll use $b = 29252$. Bob sends Alice the value $B = g^b \bmod p$. Compute $14^{29252} \bmod 36479$ to send to Alice.

   (c) Verify that when Alice computes $S = B^a \bmod p$ she gets the same answer that Bob does when he computes $S = A^b \bmod p$. This is their shared secret that no one else knows.

6. A good estimate for the area under a curve can be obtained using the Midpoint rule, which approximates the exact area using rectangles. Consider the curve $y = 1 + x^2$ on the interval $[0,1]$. Using Calculus, we can verify that the exact area is $4/3$, but even without Calculus we can approximate the area with rectangles.

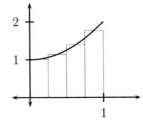

Each rectangle has width $1/4$ and touches the function at the midpoint of its top side, giving rectangle heights of $1 + (1/8)^2$, $1 + (3/8)^2$, $1 + (5/8)^2$, and $1 + (7/8)^2$.

(a) Estimate the area under the curve by computing the sum:

$$\sum_{i=0}^{3} 0.25 \left(1 + \left(\frac{1 + 2i}{8}\right)^2\right)$$

(b) Compute an expression for the area estimated by n rectangles:

$$\sum_{i=0}^{n-1} \frac{1}{n} \left(1 + \left(\frac{1 + 2i}{2n}\right)^2\right)$$

7. To win the jackpot of the Missouri lottery, a ticket holder must correctly match 6 of 40 numbered balls (in any order).

(a) Compute $\binom{40}{6}$, the number of combinations of balls that may be a winning lottery number.

(b) Each ticket holder picks two (different) sets of six numbers for the ticket and wins if either set of six is an exact match. What is the probability that an individual ticket wins the jackpot?

(c) If a person plays every month for 30 years, compute the probability of winning at least one time: $1 - \left(1 - 2/\binom{40}{6}\right)^{360}$.

8. Let $M = \begin{bmatrix} 1 & 1 \\ 1 & 0 \end{bmatrix}$.

(a) Find M^2, M^3, ..., M^{10}.

(b) Do your answers suggest a way to compute Fibonacci numbers? Find the 100th Fibonacci number.

(This method is much more efficient than the computation in Example 12.42 and probably faster than the computation in Example 12.31.)

9. Find solutions to the following equations or systems of equations. Hint: If you use Mathematica, the command you are looking for is Solve. Maple and Maxima both use solve() to find algebraic solutions. Check the help for examples and syntax.

(a) Find x, if $x^2 + x = 1$.

(b) Find x, if $x^2 + x = -1$.

(c) Find x and y.

$$4x - 3y = 5$$

$$6x + 2y = 14$$

(d) Find x, y, z, and t.

$$-2x - 2y + 3z + t = 8$$

$$-3x + 0y - 6z + t = -19$$

$$6x - 8y + 6z + 5t = 47$$

$$x + 3y - 3z - t = -9$$

10. Some equations are difficult or impossible to solve explicitly, even with software. In such situations, we often resort to numerical methods. Mathematica uses `FindRoot`, Maple uses `fsolve()`, and Maxima uses `find_root()` to find numerical solutions to equations. Here is an example where a numerical approach works well.

    Assume that I invest \$250 at the beginning of the year, \$300 at the beginning of the second quarter, \$350 at the beginning of the third quarter, and \$400 at the beginning of the fourth quarter. At the end of the year, I have \$1365 (because my investments grow). To find my (continuous) rate of return, solve this equation for r:

    $$250e^{1.0r} + 300e^{0.75r} + 350e^{0.5r} + 400e^{0.25r} = 1365.$$

11. If n is a positive number, and $g > 0$ is any "guess" for the square root of n, then a better estimate of $\sqrt{n}$ is the average of g and n/g, i.e., $\dfrac{g + n/g}{2}$. Write a function called `mysqrt` that accepts one argument, begins with an initial guess of 1.0, finds 20 new guesses, and returns the answer.

12. The Collatz conjecture states that if we start from any natural number $a_0 = n$ and form a sequence by the rule

    $$a_{i+1} = \begin{cases} a_i/2 & \text{if } a_i \text{ is even} \\ 3a_i + 1 & \text{if } a_i \text{ is odd,} \end{cases}$$

    then the sequence eventually contains the value 1. For example, starting from $a_0 = 6$, we get the sequence 6, 3, 10, 5, 16, 8, 4, 2, 1 (we reached 1 after eight steps).

    (a) Write a (recursive) function called `collatz` that accepts a single argument, n, and returns:

    - 0 if n is equal to 1
    - `1+collatz(n/2)` if n is even
    - `1+collatz(3*n+1)` if n is odd

    Thus, `collatz(n)` is the number of steps needed to go from n to 1.

    (b) Verify the values:

n	collatz(n)
1	0
2	1
6	8
27	111

Chapter 13

Getting Started with MATLAB® and Octave

This chapter introduces two of the most popular platforms for numerical computation, MATLAB and (the very similar and free) Octave.

13.1   What are MATLAB and Octave?

MATLAB (which stands for "matrix laboratory") is a programming language as well as an environment for scientific computing, published and sold by the software company MathWorks. MATLAB, commonly used for data analysis, is featured in many numerical analysis textbooks at the undergraduate and graduate level.

Octave is a free software program that has many of the same capabilities as MATLAB. It utilizes an essentially identical syntax, and a user who is competent with one program can easily transition to the other. Both share a simple, yet powerful, language that makes them excellent platforms for Linear Algebra and graphing.

A large part of what makes MATLAB and Octave so useful is that the fundamental data type they work on is the matrix. Representing matrices and vectors is simple within the language.

```
1> A = [ 1 2; 3 4]
A =

   1   2
   3   4

2> x = [ 1 3 5 7 9 ]
x =

   1   3   5   7   9
```

The operators and functions of the language operate naturally on vectors and matrices. The usual operators + - * / ^ perform matrix addition, subtraction, multiplication, division (multiplication by an inverse matrix), and exponentiation, respectively.

```
3> A / A
ans =

   1   0
   0   1

4> A ^ 2
ans =

    7   10
   15   22
```

In addition to matrix operations, MATLAB and Octave have a second set of coordinate-wise operators. This makes it convenient to do calculations with whole lists of numbers simultaneously. Matrix (and vector) addition and subtraction already operate coordinate-wise, of course. The other coordinate-wise operators are preceded with a dot: .* ./ .^

```
1> x = [ 1 2 3 5 7 9 ]
x =

   1   2   3   5   7   9

2> x .^ 2
ans =

   1    4    9   25   49   81
```

It is also permissible to add a constant to a vector or matrix (in which case the constant is added to every coordinate).

```
3> x + 1
```

```
ans =

   2   3   4   6   8   10
```

Most of the built-in functions of MATLAB and Octave operate on the coordinates of a vector or matrix. Thus, `abs()`, `sin()`, `cos()`, `tan()`, and `log()` perform the absolute value, sine, cosine, tangent, and natural logarithm, respectively, of each entry of a matrix.

If you need to know more about a command or operator, the `help` command gives simple help. More elaborate documentation is available with the `doc` command.

```
1> help cos
 -- Mapping Function:  cos (X)
     Compute the cosine of each element of X.

cos is a built-in mapper function

Additional help for built-in functions and operators is
available in the online version of the manual.  Use the command
'doc <topic>' to search the manual index.

2> doc cos
...
```

13.2   How to explore Linear Algebra

MATLAB and Octave are nice for exploring the kinds of ideas that mathematics students encounter in Linear or Matrix Algebra. Facilities to solve linear systems, to decompose matrices, and to compute determinants, ranks, and eigenvectors/eigenvalues are all part of the language.

For generating new matrices, the `zeros()`, `ones()`, `eye()`, and `rand()` commands are convenient. All of these can take one or two arguments. With a single argument they create an $n \times n$ square matrix, and with two arguments they create an $m \times n$ matrix. With them, we can make matrices with all zeros, all ones, ones on the main diagonal (an identity matrix), or random values in each entry.

Example 13.1. We generate a couple of matrices.

```
1> z = zeros(2,3)
z =

   0   0   0
   0   0   0

2> I = eye(3)
I =

   1   0   0
   0   1   0
   0   0   1
```

Slices of matrices

MATLAB and Octave make it easy to get at the submatrices of a matrix, including the individual entries, rows, or columns, by taking "slices" of the matrix using ranges or vectors for coordinates. For example, `A(1:2, 2:3)` denotes the submatrix of A formed from the first two rows (i.e., the range from 1 to 2) and the second two columns (from 2 to 3). A single colon can also serve as a (full) range, so `A(2,:)` denotes the second row of A.

```
1> A = [ 1,2,3; 4,5,6; 7,8,9 ]
A =

   1   2   3
   4   5   6
   7   8   9

2> A(3,3)
ans =

   9

3> A(2,:)
ans =

   4   5   6

4> A(1:2, 2:3)
ans =

   2   3
   5   6
```

A slice can also be determined by a vector, in which case only the rows (or columns) in the vector are part of the slice, and they appear in the order they are specified in the vector. We can use this to grab very selective parts of a matrix, or even to do matrix manipulations (like swapping two rows of a matrix).

```
5> A([1,3], [2,1])
ans =

   2   1
   8   7

6> A([1,2], :) = A([2,1], :)
A =

   4   5   6
   1   2   3
   7   8   9
```

Determinant and rank

The `det()` command returns the determinant of a matrix, and `rank()` returns the rank. The apostrophe (') operator returns the adjoint of a matrix, which is the (complex conjugate) of the transpose.

Example 13.2. We find the row rank, the column rank (i.e., rank of the transpose), and the determinant of a matrix.

```
1> A = [ 1,2,3; 4,5,6; 7,8,9 ]
A =

   1   2   3
   4   5   6
   7   8   9

2> rank(A)
ans =  2
3> rank(A')
ans =  2
4> det(A)
ans = 0
```

Solving linear systems

MATLAB and Octave use the backslash operator (\) for solving linear systems. Consider the system of linear equations

$$x - 3y - 7z = 4$$

$$-2x - 5y = 3$$

$$4x - 1y - 2z = 5.$$

This system can be written as a matrix equation that we can solve.

$$\begin{bmatrix} 1 & -3 & -7 \\ -2 & -5 & 0 \\ 4 & -1 & -2 \end{bmatrix} \begin{bmatrix} x \\ y \\ z \end{bmatrix} = \begin{bmatrix} 4 \\ 3 \\ 5 \end{bmatrix}$$

In MATLAB or Octave, the calculation would look like this.

```
1> C = [1,-3,-7 ; -2,-5,0 ; 4,-1,-2 ]
C =

   1   -3   -7
  -2   -5    0
   4   -1   -2

2> v = [ 4; 3; 5 ]
v =

   4
   3
   5
```

```
3> u = C \ v
u =

   1
  -1
   0
```

So, the solution to the system is $x = 1$, $y = -1$, and $z = 0$. We can verify this solution by multiplying.

```
4> C * u
ans =

   4
   3
   5
```

Matrix forms and decompositions

MATLAB and Octave can row reduce a matrix, find eigenvalues and eigenvectors, and compute standard matrix decompositions like the LU (lower triangular, upper triangular) decomposition and the singular value decomposition. Each command has a help entry, but here is a brief example of each.

Example 13.3. We compute the reduced row echelon form of A.

```
1>  A = [ 1,2,3; 4,5,6; 7,8,9 ]
A =

   1   2   3
   4   5   6
   7   8   9
```

```
2> rref(A)
ans =

   1.00000   0.00000  -1.00000
   0.00000   1.00000   2.00000
   0.00000   0.00000   0.00000
```

Example 13.4. We find the eigenvalues and eigenvectors of A. The column vectors of V are eigenvectors for A, and the diagonal entries of D are the corresponding eigenvalues.

```
3> [V,D] = eig(A)
V =

  -0.231971  -0.785830   0.408248
  -0.525322  -0.086751  -0.816497
  -0.818673   0.612328   0.408248
```

D =

```
    16.11684      0.00000      0.00000
     0.00000     -1.11684      0.00000
     0.00000      0.00000     -0.00000
```

Example 13.5. We decompose A into lower and upper triangular parts, L and U (with a permutation matrix P). Notice that $PA = LU$.

```
4> [L,U,P] = lu(A)
L =
```

```
     1.00000      0.00000      0.00000
     0.14286      1.00000      0.00000
     0.57143      0.50000      1.00000
```

U =

```
     7.00000      8.00000      9.00000
     0.00000      0.85714      1.71429
     0.00000      0.00000     -0.00000
```

P =

```
     0     0     1
     1     0     0
     0     1     0
```

Example 13.6. We generate a singular value decomposition for A. Notice that $A = USV'$.

```
5> [U,S,V] = svd(A)
U =
```

```
    -0.21484      0.88723      0.40825
    -0.52059      0.24964     -0.81650
    -0.82634     -0.38794      0.40825
```

S =

```
    16.84810      0.00000      0.00000
     0.00000      1.06837      0.00000
     0.00000      0.00000      0.00000
```

V =

```
    -0.479671    -0.776691    -0.408248
    -0.572368    -0.075686     0.816497
    -0.665064     0.625318    -0.408248
```

```
6> U*S*V'
ans =

   1.0000   2.0000   3.0000
   4.0000   5.0000   6.0000
   7.0000   8.0000   9.0000
```

13.3   How to plot a curve in two dimensions

It is possible to construct elegant and elaborate plots in MATLAB and Octave with relative ease. The same general principle applies: If you can describe the points of a plot, then you can plot it.

All 2D plotting is executed via the `plot()` command. In its simplest form, the `plot()` command requires two (vector) arguments that provide the (x, y) points of the graph. The coordinate-wise behavior of functions, together with a few syntactical shortcuts, make it easy to generate these vectors.

One shortcut makes it simple to create a vector of evenly spaced values. The syntax is [`low` : `step` : `high`]. For example, to create a vector containing values from 0 to 2 that are 0.25 apart:

```
1> x = [ 0 : 0.25 : 2 ]
x =

 Columns 1 through 6:

  0.00000   0.25000   0.50000   0.75000   1.00000   1.25000

 Columns 7 through 9:

  1.50000   1.75000   2.00000
```

The vectors used for graphing often contain many entries; you may have graphs with 1000 or more points, more than can be conveniently displayed as a list on the screen. Terminate any line with a semicolon to suppress the output of its calculation.

Using this syntax, it's easy to compute vectors for the (x, y) coordinates of a sine graph.

```
1> x = [ 0 : 0.01 : 2*pi ];
2> y = sin(x);
3> plot(x,y)
```

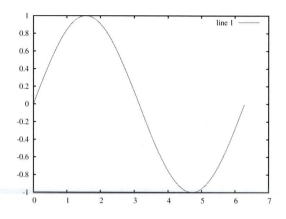

How to plot a parameterized curve

Because plots are specified simply as lists of (x, y) coordinates, there is no requirement that those points are the graph of a function (i.e., pass a vertical line test). In fact, the same method that is used to plot a function will also plot a parameterized curve.

Example 13.7. To plot a circle, create vectors containing the x and y coordinates.

```
1> t = [ 0 : 0.01 : 2*pi ];
2> x = cos(t);
3> y = sin(t);
4> plot(x,y)
```

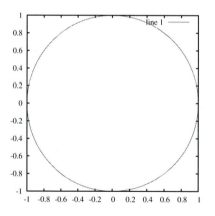

Plots in polar coordinates can also be constructed. Since graphs in polar coordinates are also a kind of parameterized plot (where x and y are functions of r and θ), they are generated in essentially the same way. Simply compute the x values and y values as lists for the `plot()` command.

Example 13.8. We graph a 3-petal rose, $r = \cos(3\theta)$.

```
1> t = [ 0 : 0.01 : 2*pi ];
2> r = cos(3*t);
3> x = r .* cos(t);
4> y = r .* sin(t);
5> plot(x,y)
```

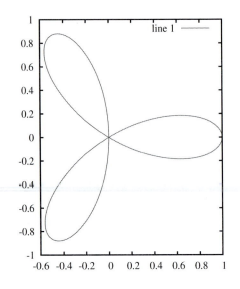

13.4   How to plot a surface in three dimensions

In a spirit similar to the way 2D plotting is accomplished in MATLAB and Octave, 3D plots are formed by generating all of the individual points (x, y, z) of the graph.

Plotting a surface on a rectangular domain

Most often, we desire 3D plots on some rectangular domain. For example, we may want a plot of $z = xy$ over the (x, y) region $[0, 2] \times [0, 1]$. The function called `meshgrid()` makes it easy to generate matrices containing the x- and y-coordinates, respectively, of a grid of points in this region.

```
1> x = [ 0 : 0.2 : 2 ];
2> y = [ 0 : 0.1 : 1 ];
3> [xx,yy] = meshgrid(x,y);
```

Here the matrices xx and yy refer to grid points on the rectangle $[0, 2] \times [0, 1]$. The grid points step by 0.2 in the x direction and 0.1 in the y direction, because these are the steps specified by the input vectors x and y. A picture of this grid looks like:

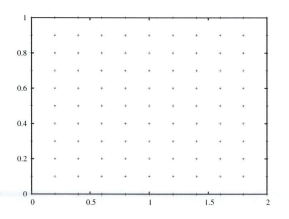

Once the meshgrid is formed, plotting the surface is simply a matter of evaluating the z-coordinates at each grid point and passing all of the values to the `mesh()` function, which draws each of the points of the surface and connects neighboring points with lines. For example, if $z = xy$, we would generate the plot this way:

```
4> zz = xx .* yy;
5> mesh(xx,yy,zz)
```

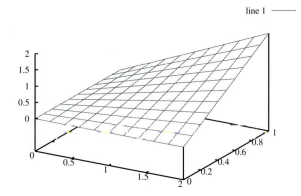

Plotting a surface on a polar domain

To generate a plot over a region described using polar coordinates, first generate a meshgrid of the polar rectangle in (r, θ) and then translate that grid into (x, y) coordinates.

For example, to plot $z = xy$ over the part of the unit disk lying in the first quadrant, i.e., over $(r, \theta) \in [0, 1] \times [0, \pi/2]$:

```
1> r = [ 0 : 0.1 : 1 ];
2> t = [ 0 : 0.2 : pi/2 ];
3> [rr,tt] = meshgrid(r,t);
4> xx = rr .* cos(tt);
5> yy = rr .* sin(tt);
```

A picture of the translated grid looks like:

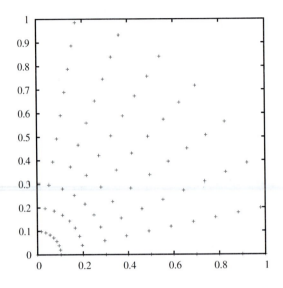

Finally, define the z-coordinates and pass everything to the `mesh()` function for plotting.

```
6> zz = xx .* yy;
7> mesh(xx,yy,zz);
```

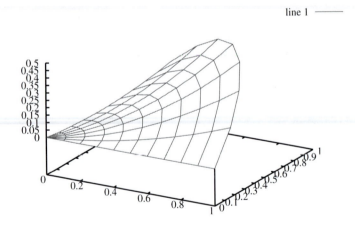

Plotting a surface from a parameterization

Because the `mesh()` function accepts matrices of (x, y, z) coordinates to plot, there is essentially no difference between plots of surfaces defined as functions $f(x, y)$ and surfaces defined by a parameterization.

As with parameterized curves, the plotting method for parameterized surfaces is simply to define the x-, y-, and z-coordinates in terms of the parameterization and pass them to the plotting function; for surfaces, this is `mesh()`.

Example 13.9. The unit sphere centered at the point $(0, 0, 1)$ is given by the spherical equation $\rho = 2\cos(\phi)$, where $\phi \in [0, \pi/2]$ and $\theta \in [0, 2\pi]$. So it is a surface parameterized by the variables ϕ and θ. To get a plot of this, form a grid in the variables of the parameterization, evaluate the function, and translate everything to (x, y, z) coordinates according to the standard spherical coordinate conversion.

```
1> t = [ 0 : 0.1 : 2*pi ];
2> p = [ 0 : 0.1 : pi/2 ];
3> [tt,pp] = meshgrid(t,p);
4> rr = 2*cos(pp);

5> xx = rr .* sin(pp) .* cos(tt);
6> yy = rr .* sin(pp) .* sin(tt);
7> zz = rr .* cos(pp);
8> mesh(xx,yy,zz)
```

See Figure 7 in the color insert.

Other parameterizations are similarly easy. The method is always the same; it begins with a meshgrid of the parameters and ends in (x, y, z) coordinates.

Plotting a parameterized curve in 3-space

Plotting parameterized space curves is one of the few places where MATLAB and Octave require a "trick" to generate a correct output. The method, as with all graphing, comes down to creating the (x, y, z) points of the curve.

Example 13.10. Say that $z(x, y) = x^2 y + 3xy^4$, where $x = x(t) = \sin(2t)$ and $y = y(t) = \cos t$. A construction of this curve might begin:

```
1> t = [ 0 : 0.1 : 2*pi ];
2> x = sin(2*t);
3> y = cos(t);
4> z = x.^2.*y + 3*x.*y.^4;
```

At this point, it would seem natural to send the vectors x, y, z to the mesh() function for drawing. Unfortunately, mesh() works differently when passed vector arguments than when passed matrix arguments, and our discussion depends on the matrix behavior.

The workaround is simple. We just need to "matricize" our vectors. The easiest way is to do something like this:

```
5> xx = [ x; x ];
6> yy = [ y; y ];
7> zz = [ z; z ];
8> mesh(xx,yy,zz)
```

This forms three matrices, each with a second row that duplicates the first. Now the mesh() function can work, essentially drawing this curve two times, once for each row. See Figure 8 in the color insert.

Notice that we could have saved some typing if we immediately matricized t to tt and based our subsequent calculations on that. However, a variation on the vector method above gives the curve a bit of vertical thickness. This can make the curve more visible if combining it with other plots.

```
9> zz = [ z+0.1; z-0.1 ];
10> mesh(xx,yy,zz)
```

See Figure 9 in the color insert.

13.5   How to manipulate the appearance of plots

The plots formed in MATLAB and Octave can be manipulated and enhanced in many ways. Custom labels may be applied to the axes, the scale may be adjusted; and for 3D plots, the viewpoint changed, contour lines added, etc.

Titles and axes

The plot title and axis labels (including the zlabel, not shown below) can all be specified.

```
1> x = [-pi : 0.1 : 3*pi ]; plot( x, tan(x))

2> title( 'Tangent function' )
3> xlabel( 'x axis' )
4> ylabel( 'y axis' )
```

To force the axes into square (so that a unit on the y-axis has the same length on screen as a unit on the x-axis), set:

```
5> axis equal
```

The limits of the axes can be specified explicitly with the `axis()` command. Pass a single vector with two, four, or six values. A two-valued vector is interpreted as the lower and upper x-axis limit. A four-valued vector sets both the x and y limits, and a six-valued vector sets the limits for all axes.

```
5> axis( [ -pi,3*pi,-2.5,2.5 ] )
```

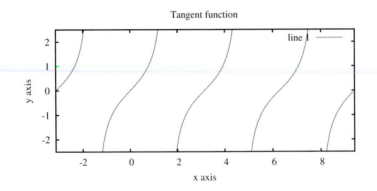

Colors and styles

An optional third argument may be supplied to the `plot()` command to alter the color and style of the plot points. This is called a "format" in the documentation, and consists of a string of options. It can be as simple as a single letter, which specifies color (r for red, g for green, etc.).

Different kinds of points may also be specified. The most common graphing method is '-', indicating points connected by lines. Other options are . * + o x, which create discrete dot, star, plus, open dot, or crossed points, respectively. Color and line styles can be combined.

```
1> x = [ 0 : 0.1 : 1 ];
2> plot( x, x, 'g-' )
3> plot( x, x, 'ro' )
```

Combining plots

There are two ways to combine multiple plots. For 2D plots, the easiest way is to pass multiple sets of arguments to the `plot()` function. Each set of three refers to a new plot. For example, to plot a quadratic and cubic function together:

```
1> x = [ 0 : 0.1 : 1 ];
2> plot( x, x.^2, '-', x, x.^3, '-' )
```

Example 13.11. In the case of multiple plots, it often makes sense to explicitly provide titles for a legend. In Octave (but not MATLAB), legend titles can be included in the format, set off by semicolons on each side.

```
1> x = [ 0 : 0.1 : 1 ];
2> plot( x, x.^2, '-;f(x)=x^2;', x, x.^3, '-;g(x)=x^3;' )
```

See Figure 10 in the color insert.

It is also worth noting that there is no requirement that the domains of each separate plot be the same. As always, plotting is simply a matter of constructing points. It might be useful to plot a long piece of a function with a short piece of tangent line.

```
1> x0 = [ -pi : 0.1 : pi ];
2> x1 = [ -1 : 0.1 : 1 ];
3> plot( x0, sin(x0), '-', x1, x1, '-' )
4> axis equal
```

The second approach to multiple plots is to create one plot, execute `hold on`, and then add a second plot. This can be a useful approach for adding extra parts to a plot in an ad hoc manner, and can also be useful when each part of a plot is complex to create.

"Hold on" is also the only way to combine 3D plots.

Example 13.12. We can add a plot of the polar domain to the plot of $z = xy$ already constructed above. Simply hold the plot, and plot another mesh where all the z-values are zero.

```
1> r = [ 0 : 0.1 : 1 ];
2> t = [ 0 : 0.2 : pi/2 ];
3> [rr,tt] = meshgrid(r,t);
4> xx = rr .* cos(tt);
5> yy = rr .* sin(tt);
6> zz = xx .* yy;
7> mesh(xx,yy,zz);
8> hold on
9> mesh(xx,yy,0*zz);
```

See Figure 11 in the color insert.

Another good use of "hold on" might be to plot a patch of tangent plane against a surface.

13.6   Other considerations

Using linearly spaced points

If you rotate some of the 3D graphs or even look closely at some of the 2D graphs, you may notice that they are slightly incomplete. There is a thin slice missing from the sphere (Example 13.9 and Figure 7 of the color pages), and the circle (Example 13.7) and rose (Example 13.8) fail to completely close, though the gaps may be too small to discern. The cause is the same in each case; it has to do with how the domain was generated.

Consider the following computation:

```
1> [ 0 : 0.5 : 1.3 ]
ans =

   0.00000   0.50000   1.00000
```

Notice that there is no value 1.3 occurring in the output. A similar behavior occurs when defining a vector [0 : 0.1 : 2*pi]. That "last point" is always missing. As an alternative to defining input vectors this way, there is a function called linspace() that returns an array of linearly spaced values between a specified beginning and end value. It works better in many plotting situations, particularly where failing to reach one end of the interval is visibly noticeable.

To get a perfect circle plotted with 99 line segments, we can use the linspace() function to return a vector of 100 evenly spaced θ values between 0 and 2π. Because both 0 and 2π will be in the resulting vector, this plot closes the small gap that the previous circle example exhibited.

```
1> t = linspace( 0, 2*pi, 100 );
2> x = cos(t);
3> y = sin(t);
4> plot(x,y)
```

How to define your own functions

It is sometimes convenient to give names to the functions used in plotting. The simplest functions are "anonymous" functions. Anonymous functions are appropriate when your function can be written as a single expression.

Example 13.13. We define two anonymous functions, $c(x) = x^3$ and $f(n,d) = \dfrac{n}{d}$.

```
1> c = @(x) x.^3;
2> c(2)
ans = 8
3> c( [2,3,4] )
ans =

    8    27    64

4> f = @(n,d) n./d;
5> f(1,3)
ans =   0.33333
```

Functions can also be defined as procedures, as in a programming language, and the output can be generated by any algorithm that you can think of and program. Here is a simple function.

```
1> function retvar = myfunc(arg1,arg2)
> retvar = arg1 .* arg2.^2;
> end
```

Note. This code example works in Octave, but MATLAB does not allow the definition of (non-anonymous) functions at the prompt. Instead, you should use an editor such as Notepad to put the definition in an "m-file," which is a text file named after the function with an extension of `.m`. A function named `myfunc()` should go in a file called `myfunc.m`, where MATLAB (or Octave) will find the definition and use it.

Note. The `pwd` command lists the "present working directory," which is usually a suitable place for saving your m-files. The `path` command lists all of the locations where the program looks for m-files.

In the function declaration above, the name of the function is declared to be "myfunc" and it takes two arguments. Within the definition of the function, the arguments are accessible as `arg1` and `arg2`. The return value will be the value assigned to `retval` when the function ultimately ends. Any values for `arg1`, `arg2`, or `retval` outside the function are irrelevant in this context.

The function above is a representation of $f(x, y) = xy^2$.

Normally, you will give more meaningful names to your functions and arguments than we have. For example, `f` would make more sense for the name of a function $f(x, y)$, with `x` and `y` (rather than `arg1` and `arg2`) for the names of the arguments and `z` for the return value.

13.7 How to learn more

The MathWorks Web site `www.mathworks.com` has extensive documentation on MATLAB, including each of the optional toolboxes. The site also contains PDF (printable) versions of the documentation.

The Octave Project Web site `www.gnu.org/software/octave` contains a comprehensive Octave manual. A printed version of their documentation, *GNU Octave Manual Version 3*, is available for purchase from Network Theory Ltd.

Exercises

1. Create a row vector that contains the values 4, 1, 3, 6, 8, 2, and 6; then use it to find the mean, median, and mode of the list. Hint: You can probably guess the names of the functions you need.

2. Determine what each command does.

   (a) `A = ones(4) - eye(4)`

(b) `A = rand(4)*10`

(c) `A = round( rand(4)*10 )`

(d) `A = round( rand(4)*10 ) + i*round( rand(4)*10 )`

(e) `A = round( (rand(4) - rand(4))*10 )`

3. Let A be defined by the command `A = diag( ones(1,4), 1 )`. Compute A, A^2, A^3, and A^4.

4. Let $A = \begin{bmatrix} 1 & -3 & -3 \\ -1 & 1 & 2 \\ 0 & -2 & 1 \end{bmatrix}$ and $B = \begin{bmatrix} 1 & -3 & 1 \\ 3 & 0 & 2 \\ 2 & 1 & 2 \end{bmatrix}$.

(a) Compute $C = AB$.

(b) Verify that `C(1:2, 1:2)` is the same matrix as `A(1:2, :) * B(:, 1:2)` (the first two rows of A times the first two columns of B).

(c) Redo parts (a) and (b) with randomly generated matrices A and B to see if the pattern always holds.

(d) Generalize the pattern/rule if you can.

5. (a) Use randomly generated matrices A to convince yourself that the determinant of AA' is always nonnegative. Is this still true if A has complex number entries?

(b) Let A be an $m \times n$ matrix, $B = AA'$, and x an $m \times 1$ column vector. Do you think $x'Bx$ can be negative? How would you explore this on the computer?

6. Find the reduced row echelon form of the matrix:

$$\begin{bmatrix} 1 & -3 & -7 & 4 \\ -2 & -5 & 0 & 3 \\ 4 & -1 & -2 & 5 \end{bmatrix}$$

Compare this result with the discussion on p. 133.

7 Solve each system of equations:

(a)

$$-2x - 4z = -8$$
$$-4x - 3y = -20$$
$$-2y = -8$$

(b)

$$1y + 2z + 1t = 23$$
$$-6x - 1y + 6z - 2t = -13$$
$$-3x + 2y + 4z - 2t = 10$$
$$-3x + 1y + 3z + 4t = 34$$

8. Verify that $PA = LU$ in Example 13.5.

9. Graph the function $y = x^3 - 3x$ on an appropriate domain. Graph the line $y = -3x$ on the same plot. What happens between the curve and the line at the origin?

10. Use the following process to graph a lemniscate (figure-eight).

  (a) Create a vector t containing values ranging from 0 to 2π.

  (b) Create a vector x containing values $\dfrac{2\cos t}{1 + \sin^2 t}$.

  (c) Create a vector y containing values $\dfrac{2\sin t \cos t}{1 + \sin^2 t}$.

  (d) Plot x vs. y.

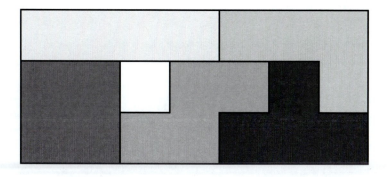

FIGURE 1: Five tetrominoes in a box (PSTricks, p. 88).

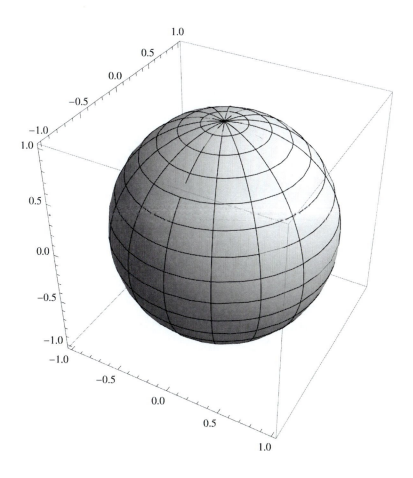

FIGURE 2: A parametric plot of a sphere (Mathematica, p. 115).

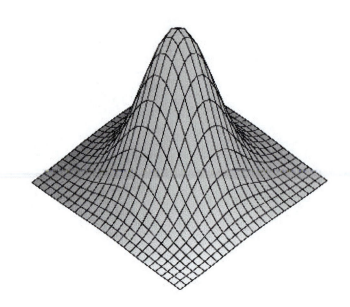

FIGURE 3: A plot of a surface (Maple, p. 117).

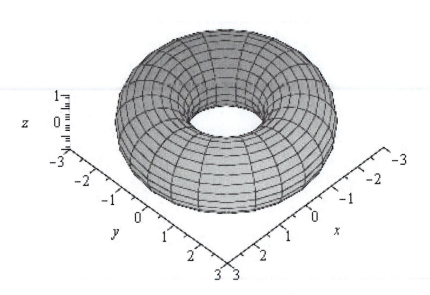

FIGURE 4: A parametric plot of a torus (Maple, p. 117).

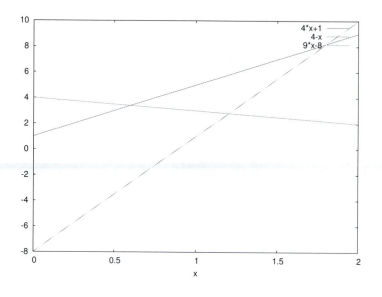

FIGURE 5: A plot of three lines (Maxima, p. 118).

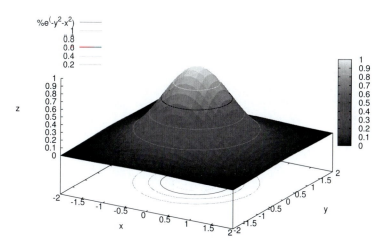

FIGURE 6: A surface plot with contour lines (Maxima, p. 118).

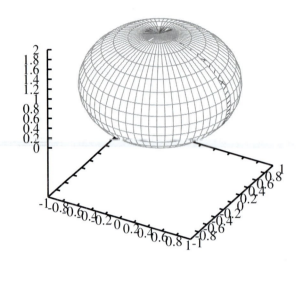

FIGURE 7: A plot of a sphere (Octave, p. 140).

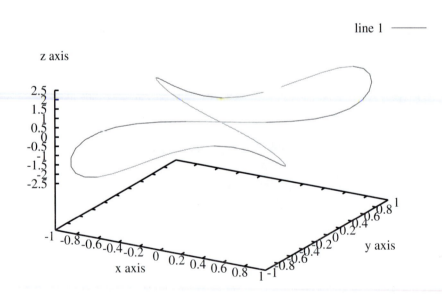

FIGURE 8: A 3-D curve (Octave, p. 141).

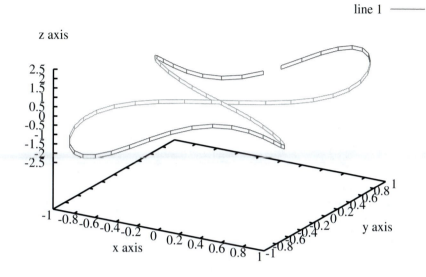

FIGURE 9: A fattened 3-D curve (Octave, p. 141).

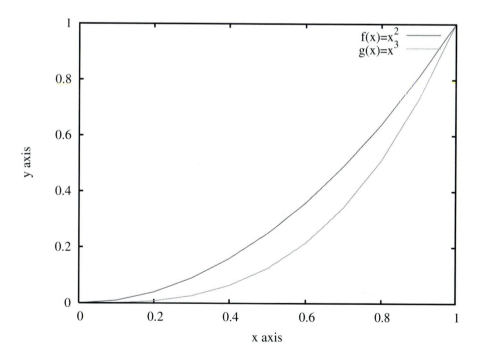

FIGURE 10: Two curves (Octave, p. 143).

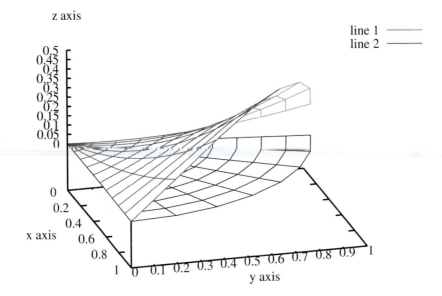

FIGURE 11: Two surfaces (Octave, p. 143).

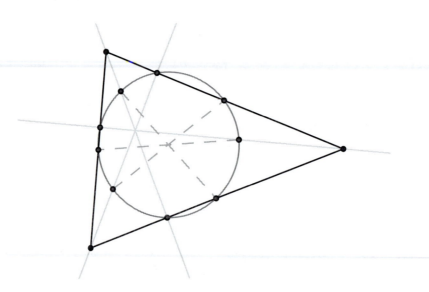

FIGURE 12: The 9-point circle (GeoGebra, p. 191).

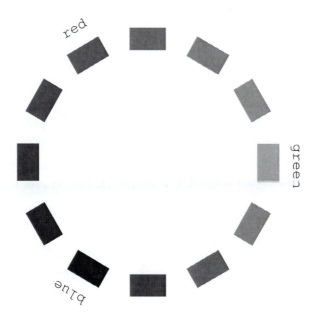

FIGURE 13: A color wheel (PostScript, p. 210).

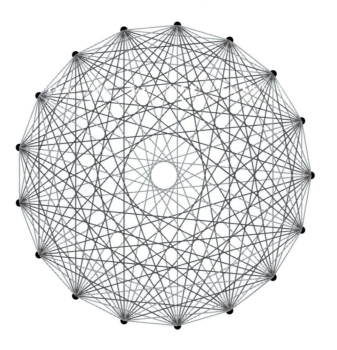

FIGURE 14: A two-colored complete graph (PostScript, p. 213).

FIGURE 15: Twelve pentominoes in a box (PostScript, p. 219).

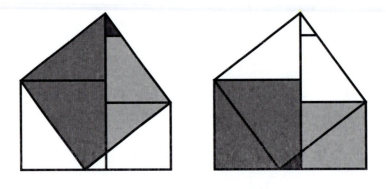

FIGURE 16: Proof of the Pythagorean Theorem (PostScript, p. 220).

Chapter 14

Getting Started with R

We take a look at R, a popular statistics package for mathematics students.

14.1 What is R?

R is a powerful free software program for doing statistics. It is similar to the S language (a popular statistical programming language) and can be used for exploring and plotting data, as well as performing statistical tests.

A mathematics student would do well to learn any of a number of statistical programs, including the commercially successful SPSS, SAS, Minitab, and S-PLUS. Still, R is a compelling platform. It is simple enough that a beginner can learn it with only a modest effort. It is powerful enough to grow with the serious student through graduate school and on to professional work. R is also free, both in cost and in a freedom sense (it has an open source license like the software discussed in Chapter 19) that we value as academics.

14.2 How to use R as a calculator

Although the major purpose of R is to work with statistical data, it can also serve as a calculator. It is not as powerful as a full computer algebra system, having no provision for imaginary numbers or purely algebraic computations. It does know basic constants like π, the usual trigonometric functions and inverses (with angles in radians), as well as the natural exponential and logarithm function, so mathematicians will feel at home.

Example 14.1. We use R as a simple calculator.

```
> pi
[1] 3.141593
> sin(pi/2)
[1] 1
> sqrt( 3^2 + 4^2 ) + log(1)*exp(0)
[1] 5
```

Note. The [1] that occurs at the left of the output technically means that the output of each calculation is a list. That does not mean much in our simple examples, because each output is a list with only one item. If the output were a list with enough items to wrap to the next line, subsequent lines would get (higher) numbers indicating how far through the list we had gone. That is, the sixth item would be preceded by [6], and so on.

R has an extensive help system containing information (with examples) about all the built-in functions and constants.

Example 14.2. We learn from the help system how to do logarithms in other bases.

```
> help(log)
...
> log(64, base=4)
[1] 3
```

Variables in R are created with the assignment operator, which historically has been the left arrow (<-). Modern versions of R also accept equals (=) for assignment, and you may find reference documents that use either symbol.

Example 14.3. We define and use a couple of variables.

```
> a = 5
> a^2
[1] 25
> b <- a
> b^2
[1] 25
```

A list of currently defined variables can be generated with ls(), and variables may be removed with the rm() function.

Example 14.4. We define and remove a variable.

```
> a = 5
> ls()
[1] "a"
> rm( a )
> a
Error: object "a" not found
```

Note. Function calls in R, such as `ls()`, always require parentheses, even when there are no arguments. This is similar to some other programming languages, notably C. Without the parentheses, R will return the definition of the function rather than evaluating it.

R works naturally with array variables, since data commonly occur in lists. The two most typical ways to create arrays in R are via the `c()` concatenation function and the `scan()` data input function.

Example 14.5. We concatenate several values together into an array a.

```
> a = c( 2,3,5,7,11 )
> a
[1]  2  3  5  7 11
```

The `scan()` function is usually more convenient for longer sets of data that can be cumbersome to enter using the concatenation function. Data may be scanned from either the keyboard or from a file. Individual data values should be separated by white space (by default), either on the same line or on adjacent lines. It is possible to specify a different delimiter for data that exists in other formats; check the help. The end of data is indicated by the first blank line or the end of file.

Note. R will prompt with the position of the next item to read. If five items have already been typed, the prompt is changed to `6:` to indicate that the sixth item is next.

Example 14.6. We read in the first ten prime numbers, as they are typed from the keyboard and then from an existing file called `primes.txt`.

```
> primes1 = scan()
1: 2 3 5 7 11
6: 13 17 19 23 29
11:
Read 10 items
> primes2 = scan( file="primes.txt" )
Read 10 items
> primes1
 [1]  2  3  5  7 11 13 17 19 23 29
> primes2
 [1]  2  3  5  7 11 13 17 19 23 29
```

Most mathematical operators in R work component-wise on arrays. For addition and subtraction this is exactly the way mathematical arrays work. Unlike mathematical arrays, R arrays can sometimes be combined even when they are not the same size. For example, an array of length 3 can be added to an array of length 6 (and the answer is an array of length 6). To make the process work, R will expand the shorter array by reusing entries that start from the left; the array $[1, 2, 3]$ would be expanded to $[1, 2, 3, 1, 2, 3]$, for example.

Example 14.7. We do some array arithmetic.

```
> a = c(1,2,3)
> b = c(5,5,5,5,5,5)
> a^2
[1] 1 4 9
> 4+a
[1] 5 6 7
> a+b
[1] 6 7 8 6 7 8
> sin( a*pi/2 )
[1]   1.000000e+00   1.224606e-16 -1.000000e+00
```

14.3 How to explore and describe data

R includes descriptive statistics and plots for summarizing data. It can compute measures of center, like mean and median, as well as measures of spread, like (sample) standard deviation and range.

Example 14.8. We compute some statistics for a set of exam scores.

```
> scores = scan()
1: 81 81 96 77
5: 95 98 73 83
9: 92 79 82 93
13: 80 86 89 60
17: 79 62 74 60
21:
Read 20 items
> range(scores)
[1] 60 98
> median(scores)
[1] 81
> mean(scores)
[1] 81
> sd(scores)
[1] 11.3555
```

There are a number of plots and charts in R for presenting or exploring data. For example, we might wonder if a set of exam scores is normally distributed. A stem and leaf plot can help us decide this, and R can generate one.

Example 14.9. We explore a group of exam scores to learn about the shape of the distribution.

```
> scores = scan()
1: 81 81 96 77
5: 95 98 73 83
9: 92 79 82 93
```

```
13: 80 86 89 60
17: 79 62 74 60
21:
Read 20 items
> stem(scores)

  The decimal point is 1 digit(s) to the right of the |

  6 | 002
  7 | 34799
  8 | 0112369
  9 | 23568


> stem(scores,scale=2) # same plot stretched twice as long

  The decimal point is 1 digit(s) to the right of the |

  6 | 002
  6 |
  7 | 34
  7 | 799
  8 | 01123
  8 | 69
  9 | 23
  9 | 568
```

Other common ways to explore a set of data are through histograms or box plots. In R, the hist() function produces histograms, and the boxplot() function creates box plots.

Example 14.10. We depict the same set of exam grades with a histogram.

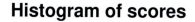

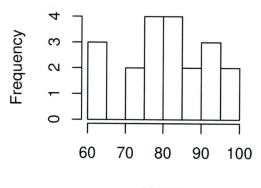

Example 14.11. We explore the same set of exam grades with a histogram that we have customized with breaks on the "fives."

```
> hist(scores,breaks=c(55,65,75,85,95,105))
```

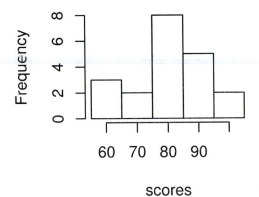

Histogram of scores

Example 14.12. Finally, we take another look at the same exam grades with a box plot.

```
> boxplot(scores)
```

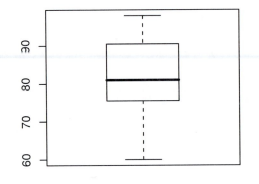

Sometimes, it is important to know if our data are normally distributed. For example, the box plot of scores above seems to suggest that those exam scores are approximately normal (and the stem and leaf plot done earlier seems to agree). There is yet another way to check, called a normal probability plot, produced by the qqnorm() function. Normally distributed data will result in a plot that looks essentially linear, and qqline() draws a reference line that we can compare against.

Example 14.13. We verify the approximate normality of our exam scores.

```
> qqnorm(scores)
> qqline(scores)
```

Normal Q–Q Plot

14.4 How to explore relationships

We can use R to learn about the relationship between two variables. For example, we may suspect that there is a relationship between the height of a person and the shoe size of the same person. Similarly, it would be unsurprising to find that people who score well on the first exam in a college class might score well on the second.

Some of the functions that summarize one set of data can work on multiple sets of data. For example, if we have data about two exams in a Calculus class, we can compare the box plots to get an idea about the relative difficulty of the exams.

Example 14.14. We look at box plots of exam scores from two exams.

```
> exam1 = scan()
1: 89 64 72 91 74
6: 94 99 78 67 85
11: 71 97 80 72 79
16: 91 86 92 91 86
21: 64 88 89
24:
Read 23 items
> exam2 = scan()
1: 73 48 86 88 69
6: 89 92 64 39 46
11: 68 81 75 70 75
16: 61 55 85 73 66
```

```
21: 18 68 83
24:
Read 23 items
> boxplot( exam1, exam2 )
```

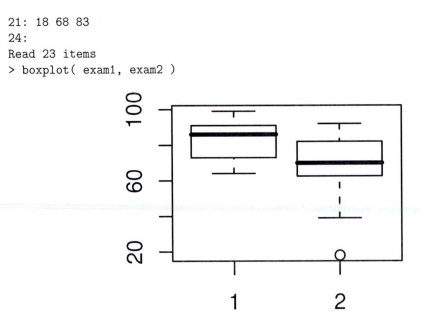

It would seem that the second exam was generally harder than the first. The median is visibly lower on the second exam, and the range of values extends much lower than the first.

To see if individual students score consistently from one exam to the next, we can use a scatter plot, generated by the plot() function. A quick look tells us immediately if two variables (like exam score on exam1 and exam score on exam2) are related by plotting them against each other.

Example 14.15. We construct a scatter plot of exam1 and exam2 scores.

```
> plot( exam1, exam2 )
```

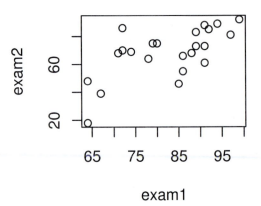

If there is a trend in these scores, it seems to be weak, but there does seem to be some pattern. Students who scored lower on exam 1 also appear to score lower on exam 2, at least in a broad sense. It looks like there could potentially be a weak linear relationship between exam scores on exam 1 and corresponding scores on exam 2. R can help us quantify that.

The `cor()` function computes correlation coefficients. It can compute them in different ways, including both Pearson's method (the most common) and Spearman's method.

Example 14.16. We check the relationship between exam1 and exam2 with Pearson's correlation.

```
> cor( exam1, exam2 )
[1] 0.6409561
```

According to R, there is some weak evidence of a linear relationship here (a stronger relationship would have a correlation coefficient closer to 1). R can also compute a line of best fit, although scores on exam 1 will be at best marginal predictors for scores on the subsequent exam. In any event, it is the `lm()` function that computes linear models of this sort. The full syntax is `lm( y ~ x )`, where y is the dependent variable and x is the independent or predictor variable.

Often, we would also like to see the line of best fit displayed with our data, and `abline()` conveniently adds lines (of any type, including linear models) to a plot.

Example 14.17. We compute a line of best fit for data from exams 1 and 2, then add it to the scatter plot.

```
> lm( exam2 ~ exam1 )

Call:
lm(formula = exam2 ~ exam1)

Coefficients:
(Intercept)        exam1
    -21.239        1.085

> plot( exam1, exam2 )
> abline( lm( exam2 ~ exam1 ))
```

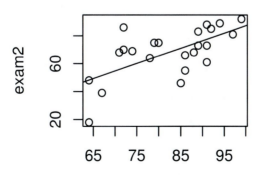

14.5   How to test hypotheses

A battery of statistical tests is built into R. Among the tests that R can perform are the familiar Student's t-test (with the `t.test()` function), the p-test of proportion (the `p.test()` function), and the χ^2-test (the `chisq.test()` function). If you have a statistical test to perform, chances are it has been written for R.

Let's consider a situation for which we might use a t-test. A stern-faced mathematics professor declares that grades in his Calculus class are normally distributed with a mean of $\mu = 75$.

We have 20 randomly selected students, and their course grades are 81, 81, 96, 77, 95, 98, 73, 83, 92, 79, 82, 93, 80, 86, 89, 60, 79, 62, 74, 60. Is the professor correct, or is the mean something different from $\mu = 75$? In R, we might test the hypothesis in the following way.

Example 14.18. We test the (claimed) class average of a math professor.

```
> scores = scan()
1: 81 81 96 77
5: 95 98 73 83
9: 92 79 82 93
13: 80 86 89 60
17: 79 62 74 60
21:
Read 20 items
> t.test( scores, mu=75 )

One Sample t-test

data:  scores
t = 2.363, df = 19, p-value - 0.02894
alternative hypothesis: true mean is not equal to 75
95 percent confidence interval:
 75.68546 86.31454
sample estimates:
mean of x
      81
```

According to R, the true mean of scores for our stern-faced professor is not $\mu = 75$. Helpfully, R tells us that we can be 95% confident that the true mean is between 75.68546 and 86.31454. Indeed, it appears that the average class grade is probably higher than 75.

By default, R computes a two-sided t-test, i.e., testing if the true mean is different from 75 (either greater or less than 75). One-sided tests are available as well. For example,

```
t.test( scores, alternative="greater", mu=75)
```

tests that the true mean is greater than 75. Similarly, setting the alternative to "less" tests that the true mean is less than 75.

14.6   How to generate table values and simulate data

Students in statistics courses usually become very familiar with the statistical tables in their textbooks. The z-table and the t-table are heavily used throughout introductory courses, and other tables become important for more advanced work. With computers, we should not need printed tables, and R contains functions that compute the values of all the common statistical tables.

The most important table is the z-table, which contains values of the normal probability distribution function. There are two functions in R that replicate the two ways this table is used: the `pnorm()` function, which takes z-values and returns p-values; and the `qnorm()` function, which does the inverse (it takes p-values and returns the corresponding z-value).

Example 14.19. We check that only about 5% of observations should fall more than 1.645 standard deviations left of the mean. Then we verify that approximately 68% of observations fall within 1 standard deviation of the mean (on either side).

```
> pnorm( -1.645 )
[1] 0.04998491
> pnorm(1) - pnorm(-1)
[1] 0.6826895
```

Example 14.20. We check that `qnorm()` performs the inverse.

```
> qnorm( 0.04998491 )
[1] -1.645
> qnorm( 0.5 )
[1] 0
```

Each of the different distributions has a similar pair of functions. The Student's t distribution has the `pt()` function (which takes t-values and returns p-values) and the `qt()` function (which takes p-values and returns t-values). There is also a pair for the χ^2 distribution, namely, `pchisq()` and `qchisq()`.

Example 14.21. We compute some values from the t- and χ^2-tables.

```
> pt( -1, df=15 )
[1] 0.1665851
> qt( 0.1665851, df=15 )
-0.9999999
> pchisq( 11.071, df=5 )
[1] 0.9500097
> qchisq( 0.95, df=5 )
[1] 11.07050
```

Sometimes, it's convenient to generate random numbers to do an experiment or to simulate a scenario. For example, we might like to generate dice rolls. A die is an example of a uniform probability distribution, since each side is equally likely to turn up. Uniformly distributed random numbers are generated in R with the `runif()` function. By default, `runif()` generates numbers between 0 and 1, although this may be customized.

Example 14.22. We simulate a roll of five dice. We do this by computing five uniformly distributed random numbers between 0 and 6, rounding them down (with the `floor()` function), and finally adding 1.

```
> floor( runif(5, max=6 )) + 1
[1] 4 6 2 6 4
```

Generating random data from other distributions is as easy as choosing the appropriate function in R. Normally distributed random data sets are generated by the `rnorm()` function, for example.

Example 14.23. We simulate a random sample of 20 items from a normally distributed population that has mean 50 and standard deviation 8.

```
> rnorm(20, mean=50, sd=8)
 [1] 46.76492 56.54318 38.80065 40.58243 46.18913 31.09232
 [7] 29.51878 38.60033 51.89929 45.93668 60.25701 49.84637
[13] 43.12018 46.71614 42.53040 53.01289 61.80566 62.21275
[19] 48.14231 36.18348
```

14.7   How to make a plot ready to print

All of the plots that R generates can be saved into files. To create a PDF of a graphic, use the `pdf()` function before your plot. To create a PNG (Portable Network Graphics) image, use the `png()` function before your plot. In either case, the `dev.off()` function closes the file when you are finished plotting.

Example 14.24. We create a 4 inch × 4 inch PDF and a 400 pixel × 400 pixel PNG of a box plot.

```
> scores <- scan()
1: 98.0 96.0 95.0 93.0 92.0
6: 89.0 86.0 83.0 82.0 81.0 81.0 80.0
13: 79.0 79.0 77.0 74.0 73.0
18: 62.0 60.0 60.0
21:
Read 20 items
> # A PDF that measures 4 inches by 4 inches
> pdf( file="scores-box.pdf", width=4, height=4 )
> boxplot( scores )
> dev.off()
null device
          1
> # A PNG that measures 400 pixels by 400 pixels
> png( file="scores-box.png", width=400, height=400 )
> boxplot( scores )
> dev.off()
null device
          1
```

14.8   How to learn more

The primary Web site for the R project is `www.r-project.org`. The project site contains copies of R for downloading, and there is a collection of online documentation and tutorials for R. For those who prefer hard-copy resources, there are links to dozens of books related to R. Since R is so similar to the S language, books on S may also be useful (and the bibliography on the R project site consequently contains many S books as well).

Exercises

1. Put the list of values 7, 5, 9, 2, 1, 8, 4, 2, 4, 8 into a variable a.

   (a) Find the mean, median, and sample standard deviation of a.

   (b) Sort the array a.

2. Put the following values into a file (using Notepad or some other suitable editor), and read the file into a variable b.

9	9	5	6	9	2	5
1	9	9	1	4	8	10
4	5	4	1	8	2	5
8	2	9	4	1	2	6
3	2	9	7	6	4	6

   (a) Generate a five number summary of b.

   (b) Create a box plot of b.

   (c) Create a stem and leaf plot of b.

   (d) Create a normal probability plot of b.

   (e) Do you think this data set is normally distributed?

3. Consider the table of data

x	2	8	13	10	12	12	8	16	5	5	14
y	4	16	18	13	19	16	8	5	19	6	7

   (a) Draw a scatter plot of data points (x, y).

   (b) Does the data set appear to have a strong or weak linear relationship?

   (c) Would you predict a correlation near -1, 0, or $+1$? Verify your intuition by calculating the correlation coefficient.

   (d) Compute a line of best fit for the data (that is, a linear model).

   (e) Add the line of best fit to the scatter plot using `abline()`.

   (f) Is the line a good predictor line?

4. Light bulb life for 60-watt bulbs is reported on the box to be 800 hours. A random sampling of 25 bulbs yields the following bulb lifetimes.

$$
\begin{array}{ccccc}
719 & 729 & 737 & 741 & 743 \\
753 & 763 & 764 & 765 & 767 \\
772 & 774 & 774 & 776 & 784 \\
786 & 801 & 807 & 809 & 810 \\
812 & 813 & 825 & 828 & 866
\end{array}
$$

(a) Verify that the tested bulb lifetimes appear to be approximately normally distributed (via `stem()`, `hist()`, `qqnorm()`, or similar commands) to confirm that the assumptions of the t-test are met.

(b) Use the help system to find the arguments to `t.test()` for specifying a one-sided alternative hypothesis.

(c) Test the claim that average light bulb life is at least 800 hours. Hint: The alternative to this claim is that bulb life is less than 800 hours.

5. Create an expression that uses the random number generator to simulate a 2d10 dice roll (the sum of the spots on two 10-sided dice), suitable for use in a role-playing game. Hint: Figure out how to generate the list of spots first, then use `sum()` around the expression to add them.

Chapter 15

Getting Started with HTML

The purpose of this chapter is to help you begin using HTML, a markup language with which you can produce pages for display on the World Wide Web. This chapter contains simple examples for you to try. It covers the basics, enough to get you started. To learn more about HTML, you may want to consult the sources listed at the end of the chapter and in the references cited.

15.1 What is HTML?

HTML (Hypertext Markup Language) is a document markup and hyperlink specification language. In a markup language, the layout of a page is specified by commands written by the author into the page. In the case of HTML, a Web browser responds to the commands and shows the pages. As a hyperlink specification language, HTML can be used to link to other documents on the Internet.

HTML was introduced by Timothy Berners-Lee at the Conseil Européenne pour la Recherche Nucléaire (CERN). The original goal was to provide a communication system for particle physicists to exchange information. The HTML format has proven popular for computer users in general, and is now the standard markup language for the World Wide Web.

Element	Use	Attributes
html	defines document	
head	defines head	
body	defines body	bgcolor, text, link, alink, vlink
title	defines title	
strong	boldfaces text	
em	emphasizes text	
center	centers text	
hn $(1 \leq n \leq 6)$	heading, subheading, etc.	
br	line break	
p	new paragraph	
ul	unordered list	
ol	ordered list	
li	list item	
object	image, document, etc.	data, type, width, height
img	image	src, width, height, alt
a	link to page, file, etc.	href

TABLE 15.1: Basic HTML elements.

15.2   How to create a simple Web page

In this section, we describe how to create a minimal file, modify text, organize text, and create lists.

A minimal page

Certain elements are necessary in every HTML document. The four basic elements are:

- html (tells the Web browser that you are writing HTML)

- head (contains information about your document that doesn't appear on the page—similar to the preamble of a LATEX document)

- title (sets the title of your page—appears in the title bar of the Web browser and in Internet search engines, such as Google)

- body (contains the material that appears on your page)

Elements are given by pairs of *tags*, one to open the element and one to close it. For example, the html element is opened with <html> and closed with </html>. Tags are nested, so that if a new element is opened within the tags for another element, then it is closed within those same tags. Some elements may be opened and closed with a single tag, e.g., <img ... />. In addition, some elements can have *attributes*, which are added commands that tailor them for specific situations. The standard practice is to use lowercase characters in HTML tags and tag attributes.

Table 15.1 displays some basic HTML elements, most of which we will discuss in this chapter.

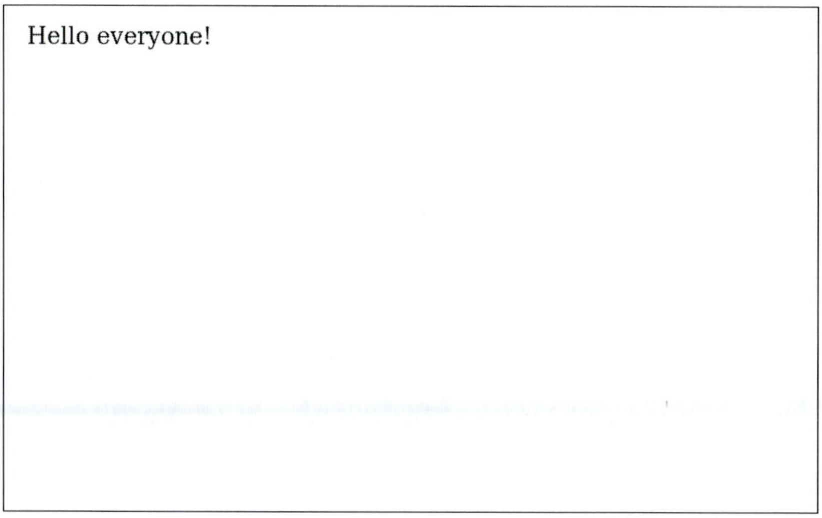

Hello everyone!

FIGURE 15.1: A simple Web page.

Note. You can create the examples in this chapter with a simple text editor such as Notepad. After creating a file, you should save it with the extension `.html` or `.htm`.

Example 15.1. We create a minimal Web page.

```
<html>

<head>

<title>My Page</title>

</head>

<body>

<p>Hello everyone!</p>

</body>

</html>
```

You can open your file (saved as, say, `example.html`) with a Web browser. The displayed page is shown in Figure 15.1.

Modifying text

Sometimes we wish to change the appearance of certain text on a Web page. For instance, we may wish to put text in italics or boldface. Italics or emphasized text is obtained with the `em` element, and boldface with the `strong` element.

- em

- strong

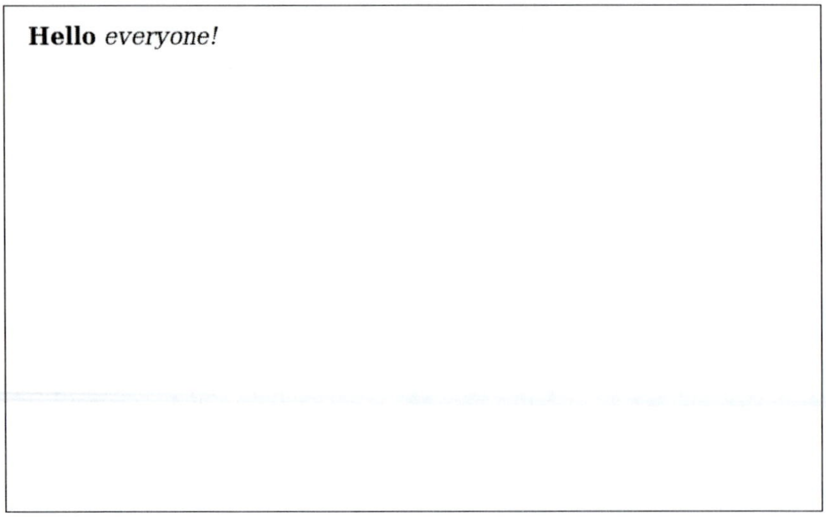

Hello *everyone!*

FIGURE 15.2: A Web page with modified text.

Example 15.2. We modify the text in the Web page of Example 15.1.

```
<html>

<head>

<title>My Page</title>

</head>

<body>

<p><strong>Hello</strong> <em>everyone!</em></p>

</body>

</html>
```

The Web page is shown in Figure 15.2.

Note. You can include special symbols in your Web page with the construction `&symbolname;`. For example, the string `&pi;` produces the symbol π. Subscripts are produced with `<sub>` and `</sub>` tags, superscripts with `<sup>` and `</sup>` tags.

Organizing text

You can organize text in an HTML document in a variety of ways. Paragraphs are produced with the p element. Centering is produced with the `center` element.

- p
- center

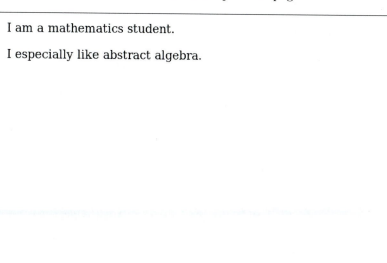

FIGURE 15.3: A Web page with paragraphs.

Example 15.3. We make a simple Web page with paragraphs.

```
<html>

<head>

<title>My Page</title>

</head>

<body>

<p>I am a mathematics student.</p>

<p>I especially like abstract algebra.</p>

</body>

</html>
```

The Web page is shown in Figure 15.3.

Note. In HTML, extra spaces and line breaks have no effect on the appearance of a page.

Making lists

Enumerated lists are made with the `ol` (ordered list) element. Bulleted lists are made with the `ul` (unordered list) element.

- `ol`

- `ul`

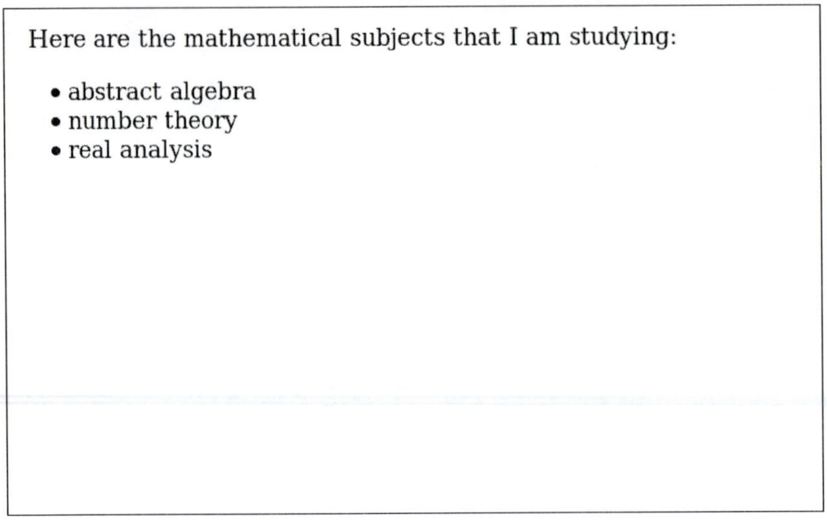

Here are the mathematical subjects that I am studying:

- abstract algebra
- number theory
- real analysis

FIGURE 15.4: A Web page with a list.

Example 15.4. We make a Web page with a list.

```
<html>

<head>

<title>My Page</title>

</head>

<body>

<p>Here are the mathematical subjects that I am studying:</p>

<ul>
   <li> abstract algebra </li>
   <li> number theory </li>
   <li> real analysis </li>
</ul>

</body>

</html>
```

The Web page is shown in Figure 15.4.

15.3   How to add images to your Web pages

You can add images to your Web page with the `img` tag.

* `img`

Note. In fact, there exists a more general tag, `object`, with which you can add images, sound, documents, etc.

Example 15.5. We make a Web page containing an image. (This image must be available in the same folder as your Web page document. You can create it yourself or copy it from another source.)

```
<html>

<head>

<title>My Page</title>

</head>

<body>

<p>Here is a complete graph with 17 vertices.</p>

<center>
<img src="completegraphorder17.gif" width="300" height="300"/>
</center>

</body>

</html>
```

The Web page is shown in Figure 15.5.

Note. Attributes (the source, width, and height in this example) should be enclosed in quotation marks.

Some commonly used image formats for displaying mathematical images on Web pages are GIF (Graphics Interchange Format) and PNG (Portable Network Graphics). Photographs are often saved in JPG (Joint Photographic Experts Group) format.

15.4   How to add links to your Web pages

A *link* (short for *hyperlink*) is a way to navigate from one Web page to another or to download a file. Links are added with the `a` element.

* `a`

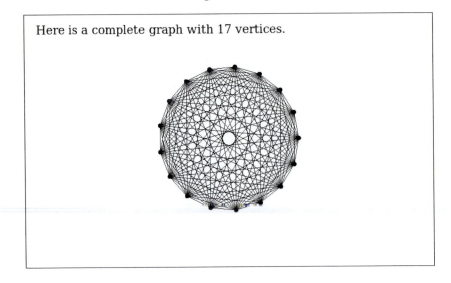

Here is a complete graph with 17 vertices.

FIGURE 15.5: A Web page with an image.

To link to a new address, type

```
<a href="http://address">text</a>
```

where `address` is the address of the link and `text` is what viewers see and "click on" to go to the linked page.

To link to a document (such as a PDF file), type

```
<a href="yourfile.pdf">text</a>
```

where `yourfile.pdf` is the file you want users to access and `text` is what viewers see and "click on" to get the file.

Note. HTTP stands for "Hypertext Transfer Protocol," a scheme for telling the Web browser how to open Web pages. This protocol is an example of a *scheme*. There are schemes for pages, documents, etc. A scheme together with an address is called a URL (Uniform Resource Locator).

Example 15.6. We create a Web page with links.

```
<html>

<head>

<title>My Page</title>

</head>

<body>

<p>Check out the
<a href="http://www.maa.org/">
Mathematical Association of America</a>.</p>
```

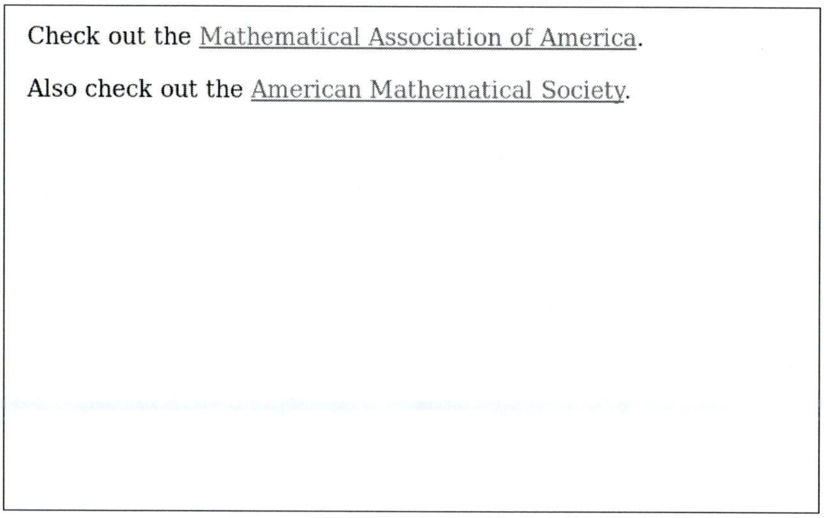

Check out the <u>Mathematical Association of America</u>.

Also check out the <u>American Mathematical Society</u>.

FIGURE 15.6: A Web page with links.

```
<p>Also check out the
<a href="http://www.ams.org/">
American Mathematical Society</a>.</p>

</body>

</html>
```

The Web page is shown in Figure 15.6.

Note. To use an image as a link, put its img tag between the a tags. The construction is

```
<a href="http://address"><img src="image.gif" width="x" height="y"/></a>.
```

15.5   How to design your Web pages

The most important aspect of Web page design is CONTENT. You should present something meaningful. Following content, the next most important consideration is CLARITY. Make your pages easy to read and navigate.

CONTENT + CLARITY

You should organize your Web pages so that it's obvious to users how to use them and find information. One popular and logical way to organize your Web pages is to follow a "top-down" directory ordering. A main page has links to some secondary pages, the secondary pages may have links to further pages, and so on. The organization is indicated in the following diagram.

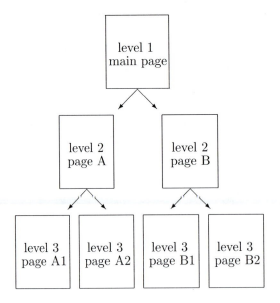

You might want to experiment by making a Web page showcasing your mathematical interests, including topics you are studying and problems you are trying to solve.

15.6   How to learn more

Many features of HTML are not discussed in this introduction, such as style sheets and the XML and XHTML generalizations. Here are some resources for you to investigate to learn more.

A good introductory book on HTML is [8]. Also, for basics, please look at the sites

> `archive.ncsa.uiuc.edu/General/Internet/WWW/HTMLPrimerAll.html`

and

> `www.utoronto.ca/webdocs/HTMLdocs/NewHTML/htmlindex.html`.

A good intermediate-level book is [40]. To learn about XML (Extensible Markup Language), see [9] and [10]. The World Wide Web Consortium (W3C) has a wealth of information on HTML and XML (including a service to check your code) at:

> `www.w3.org`

Exercises

1. What is wrong with the following HTML construction?

```
<p> This is <strong><em>bold and italics</strong></em>.</p>
```

2. What does the `<title>`...`</title>` section of a Web page contain? Where does the resulting text appear?

3. The elements `<strong>` and `<em>` are examples of semantic markup. Explain what this means and why these tags are different from `<b>` and `<i>` and `<u>` (which are examples of presentational markup).

4. Find out what the `alt` attribute does in an image link.

5. Make a Web page showcasing some of your mathematical interests.

6. Experiment with different size headings on your Web page. See Table 15.1.

7. Add a link on your Web page to one of the resources described in Chapter 7.

8. Put a link on your Web page to your résumé.

9. Put an image of a mathematical object on your Web page and describe the image.

10. Choose a theorem or a problem and make a Web page showing its proof or solution.

11. Make a Web page called "Two Proofs of the Pythagorean Theorem." Create your own pictures illustrating the proofs (you may want to use LaTeX's picture environment, PSTricks, Geometer's Sketchpad, GeoGebra, or PostScript). Construct your page so that a reader can click on an image of either proof and go to a secondary page showing that proof.

Chapter 16

Getting Started with Geometer's Sketchpad® and GeoGebra

Dynamic geometry software lets you create drawings that you can resize and reshape interactively. Dynamic drawings have a lot of potential for experimentation (particularly experimentation with geometric constructions), and they give a presenter tools that go far beyond static plots or graphics.

16.1 What are Geometer's Sketchpad and GeoGebra?

The category-defining dynamic geometry program is Geometer's Sketchpad, which is published by Key Curriculum Press. It has been in development since the 1980s and is a staple in mathematics education. Sketchpad is available for both Windows and Mac OS X. Sketchpad can save drawings in its native format and also export to "JavaSketchpad" format, which allows you to publish drawings to the Web. Once your drawings are on a Web

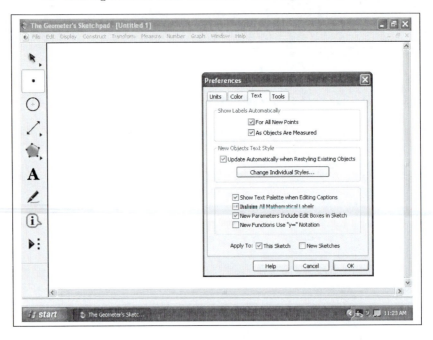

FIGURE 16.1: Geometer's Sketchpad, with the preferences dialog open.

page, anyone with a Java-enabled Web browser can view and interact with them (without needing to purchase their own copy of Geometer's Sketchpad).

GeoGebra is a similar but younger program, originally written in 2001 and available as a free download from the GeoGebra project. Like Sketchpad, GeoGebra is used for creating dynamic drawings. GeoGebra was written in Java, so it should run on any platform that supports Java, including Windows, Mac OS X, and GNU/Linux. GeoGebra can also publish drawings to the Web, where anyone can interact with them.

16.2   How to use Geometer's Sketchpad

A simple construction: Varignon's Theorem

Varignon's Theorem is a theorem of Euclidean geometry that tells us that the midpoints of the sides of any quadrilateral form the vertices of a parallelogram. It also makes a very nice construction to learn the basics of dynamic geometry software.

Figure 16.1 shows an empty drawing in Geometer's Sketchpad with the Preferences dialog window open over it. Notice the toolbar that runs down the left side of the main window. It contains tools for drawing points, circles, lines (and segments), polygons, and other objects. A menu bar also runs across the top, with menus common to all programs (like File and Edit) as well as distinctly mathematical menus (like Construct and Transform).

Before we begin our first drawing, open the Preferences dialog in your copy of Sketchpad by choosing Edit/Preferences from the menu bar. On the Text tab there is a checkbox to show labels automatically For All New Points. If you check that box and click OK, it will make the following activities easier to follow.

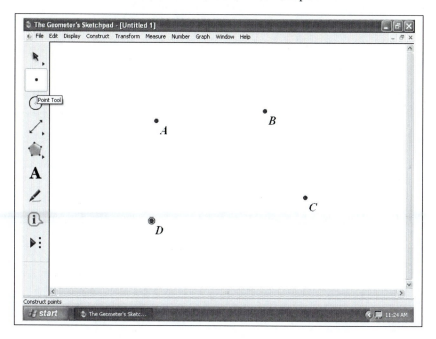

FIGURE 16.2: Geometer's Sketchpad, after drawing four points.

To begin our exploration of Varignon's Theorem, select the **Point** tool from the left toolbar by clicking the toolbar button that looks like a point. Click on the drawing area with the **Point** tool in four places to make the corners of a quadrilateral. Sketchpad will label each point with a letter (because we selected that in the **Preferences** dialog), and your result should look similar to Figure 16.2.

To create the sides of our quadrilateral, we will use the **Straightedge** tool from the toolbar. If you hold down the tool button, you will find that there are actually three tool choices: **Segment Straightedge** tool, **Ray Straightedge** tool, and **Line Straightedge** tool. We want the **Segment Straightedge** tool.

With the **Segment Straightedge** tool selected, click point A and then click point B. You will notice that as the mouse moves from A to B, Sketchpad extends a segment. This segment becomes a permanent part of the drawing when you click on B. Similarly, you can click B and C to create a second side of the quadrilateral. Finish the last two sides in the same way. Your drawing should now look something like Figure 16.3.

At this point, we have constructed a quadrilateral and it is time to add the midpoints of the sides. The first thing we need to do is change tools. There is a tool at the top of the toolbar that looks like an arrow pointer. Technically, it is the **Translation Arrow** tool. It lets you select parts of your drawing, but it also lets you pull them around (that is, "translate" them). If you pull on point A, notice that the segments that connect A to B and D automatically adjust and everything stays connected. You should find that you can pull on any of the points or segments in the diagram and still have a (possibly self-intersecting) quadrilateral.

To draw the midpoints of the sides of our quadrilateral, we will use the **Midpoint** menu item under the **Construct** menu. Depending on the items you currently have selected, the Midpoint item may be active or grayed out (that is, not available). Click anywhere in the white background to remove all existing selections, and click the segment $\overline{AB}$ to select (only) it. Then you can use **Construct/Midpoint** to create the midpoint of the segment, E.

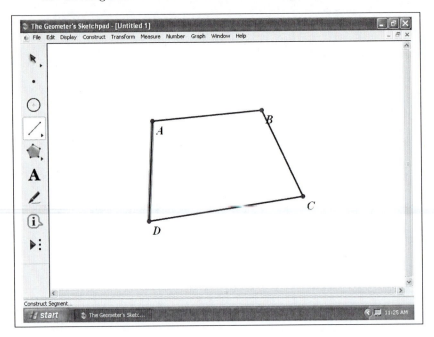

FIGURE 16.3: Geometer's Sketchpad, after connecting four points with segments.

The same process can be used to create midpoints F, G, and H.

Note. Selection is probably one of the less intuitive aspects to working with Geometer's Sketchpad. If you click on various parts of your current diagram (points, segments, and so on) you will notice that new selections do not replace old selections, but are added to them. For example, if point A is selected, and you click point B, they are now both selected (most programs would transfer the selection to B). To deselect everything in a drawing, simply click the background. You may then select only the objects you want.

Now choose the **Segment Straightedge** tool to connect the midpoints. Begin by creating the segment $\overline{EF}$. To make it easier to see the interior parallelogram, we may wish to change the color of $\overline{EF}$. To do this, right-click the segment and choose **Color** from the menu that appears. Pick a contrasting color that pleases you. Your drawing should now look something like Figure 16.4.

Note. Sketchpad will remember the color you have selected, and will use it as you create the remaining segments $\overline{FG}$, $\overline{GH}$, and $\overline{HE}$.

With the drawing finished, it is time to play. Choose the **Arrow** tool again and manipulate the drawing by pulling on different parts. You should find that no matter how you deform it, the midpoints always form the vertices of a parallelogram. Sketchpad can help us explore why this might be, and here's a hint. Use the **Line Straightedge** tool to connect point A and C. To make it stand out, right-click $\overleftrightarrow{AC}$ to choose **Dotted Line** or to select a new color. Compare to Figure 16.5.

The hint is that $\overleftrightarrow{AC}$ appears to be parallel to $\overline{EF}$, and this remains true no matter how you deform the drawing. Perhaps you will remember from geometry that $\triangle ABC$ is similar to $\triangle EBF$. For a proof, introduce coordinates for the points A, B, and C, and verify the vector relationship $\overrightarrow{AC} = 2\overrightarrow{EF}$.

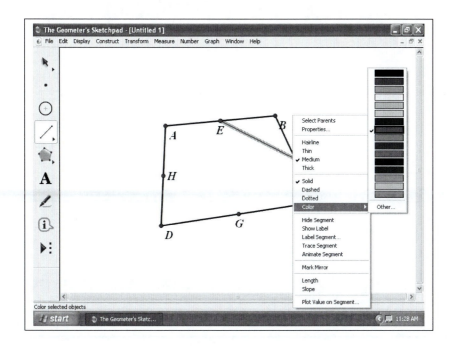

FIGURE 16.4: Geometer's Sketchpad, connecting midpoints in a new color.

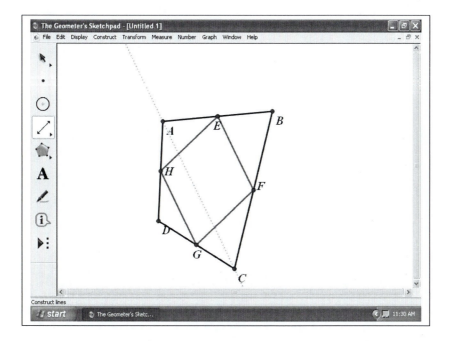

FIGURE 16.5: Geometer's Sketchpad, hint to a proof of Varignon's Theorem.

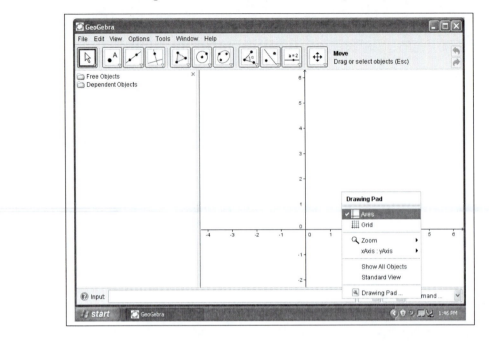

FIGURE 16.6: GeoGebra, with the Drawing Pad dialog open.

16.3   How to use GeoGebra

A simple construction: Varignon's Theorem

As with Geometer's Sketchpad, Varignon's Theorem makes a very nice construction for learning about GeoGebra.

Figure 16.6 shows an empty drawing in GeoGebra. Notice the toolbar that runs across the top of the main window. It contains tools for drawing points, circles, lines (and segments), polygons, and other objects. Most of your interaction with GeoGebra will take place through this toolbar.

Before we begin our first drawing, right-click the drawing pad and deselect the item called Axes (to turn off the x and y axes).

To begin our exploration of Varignon's Theorem, select the New Point tool from the toolbar by clicking the toolbar button that looks like a point (this should be the second button from the left). Click on the drawing area with the New Point tool in four places to make the corners of a quadrilateral. GeoGebra will label each point with a letter, and it will list each point in the left object area under "Free Objects." See Figure 16.7.

To create the sides of our quadrilateral, we will use the Segment tool from the toolbar (the third button). If you hold the small arrow on the tool button down, you will find that there are actually several tool choices. We want the Segment between Two Points tool.

With the Segment tool selected, click point A and then click point B. You will notice that as the mouse moves from A to B, GeoGebra extends a segment. This segment will become a permanent part of the drawing when you click on B. It will also be labeled and listed in the object area. Similarly, you can click B and C to create a second side of the quadrilateral. Finish the last two sides in the same way. Your drawing should now look something like Figure 16.8.

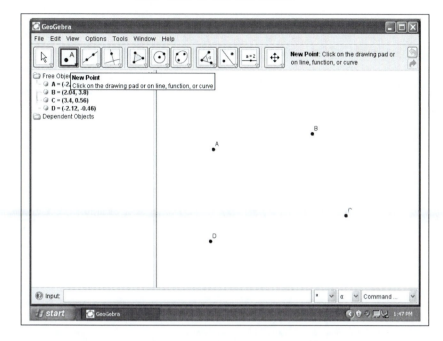

FIGURE 16.7: GeoGebra, after drawing four points.

At this point, we have constructed a quadrilateral and we should take a moment to "play" with it. Use the Move tool (the arrow pointer) and try to drag the points or segments in our drawing. You should find that you can drag any of the blue corners of the quadrilateral. If you pull on point A, notice that the segments that connect to B and D automatically adjust and everything stays connected. Similarly, if you pull on a side, the ends move along with it and other sides stay connected.

Note. Users migrating to GeoGebra from Sketchpad should know that Sketchpad works very hard to make drawing dynamic, while GeoGebra allows fewer kinds of drawing manipulation than Sketchpad allows. In GeoGebra, "free objects" can always be moved. Dependent objects are determined by their relationships with free objects. Sometimes, dependent objects can be moved (like the sides of our quadrilateral), but sometimes not, even when Sketchpad would have allowed the manipulation.

It is time to add the midpoints of the sides, and for this we use the Midpoint or Center tool. It is on the same toolbar button as the New Point tool, under the drop-down menu. With the Midpoint tool, click side a to create a midpoint E, and continue similarly with each of the remaining sides. Notice that the new points are colored black and listed as dependent objects, as shown in Figure 16.9

Now choose the Segment tool again (third toolbar button) to connect the midpoints. Connect E to F, then F to G, and so on. It should appear that $\overline{EF}$ is parallel to $\overline{HG}$.

To make the parallelogram stand out, let's change its color. Choose the Move tool, and select one segment. Then, holding down the Ctrl key (Command key on Macintosh), select the other three segments of the parallelogram. Finally, right-click on any of the selected segments and choose the Color tab, as in Figure 16.10.

With the drawing finished, it is time to play again. Choose the Move tool again and manipulate the drawing by pulling on any of the corners (the free objects). You should find that no matter how you deform it, the midpoints always form the vertices of a parallelogram. GeoGebra can help us explore why this might be, and here's a hint. Use the Line Through

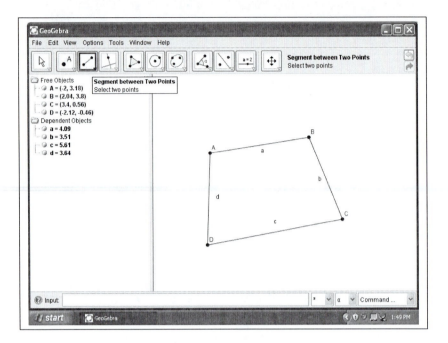

FIGURE 16.8: GeoGebra, after connecting four points with segments.

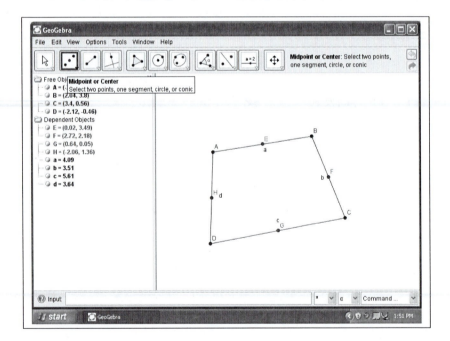

FIGURE 16.9: GeoGebra, after creating four midpoints.

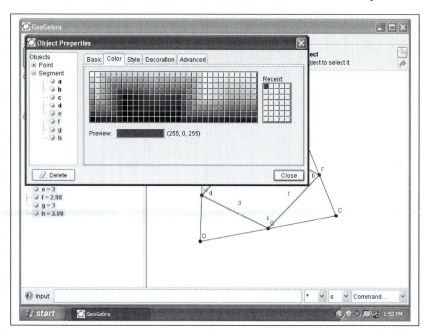

FIGURE 16.10: GeoGebra, with the Color Properties tab open.

Two Points tool to connect point A and C. To make it stand out, right-click $\overleftrightarrow{AC}$ to change the properties of the line, like Color or Style (you can have, for example, dotted or dashed lines). Compare to Figure 16.11.

The hint is that $\overleftrightarrow{AC}$ appears to be parallel to $\overline{EF}$, and this remains true no matter how you deform the drawing. Perhaps you will remember from geometry that $\triangle ABC$ is similar to $\triangle EBF$. To prove it, introduce coordinates for the points A, B, and C, and verify the vector relationship $\overrightarrow{AC} = 2\overrightarrow{EF}$.

16.4 How to do more elaborate sketches in Geometer's Sketchpad

The 9-point circle

The 9-point circle is a famous construction that can be done for any triangle. According to the construction, given any triangle there is a single circle that contains all of these points:

• The three midpoints of the sides of the triangle.

• The three intersections where the altitudes meet the sides of the triangle.

• The midpoints of the segments that connect the orthocenter to each vertex of the triangle.

It is somewhat unwieldy to describe the 9-point circle in words, but it is a terrific demonstration of geometry software.

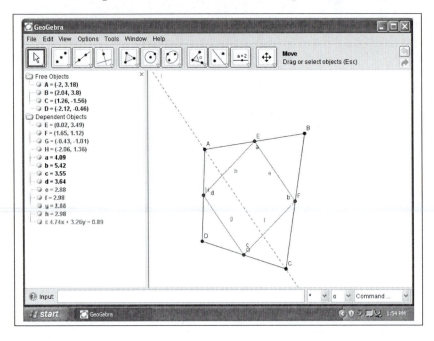

FIGURE 16.11: GeoGebra, hint to a proof of Varignon's Theorem.

To begin a construction, use the **Point** tool to create three vertices A, B, and C for a triangle. Follow with the **Segment Straightedge** tool to connect the vertices, then use the **Arrow** tool along with the **Construct/Midpoint** menu option to find the midpoints, D, E, and F of the sides of the triangle. These are the first three points of the 9-point circle. Your drawing may look something like Figure 16.12.

To find the next three points of the 9-point circle, we need to create the altitudes of the triangle. Each altitude is the line going through a vertex of the triangle and perpendicular to the opposite side. To create the first altitude, use the **Arrow** tool to select point A and segment $\overline{BC}$. Then choose **Construct/Perpendicular Line** from the **Construct** menu. The other two altitudes are formed similarly.

We need the points of intersection, so switch to the **Point** tool, and add a point where the altitudes intersect $\overline{BC}$, $\overline{AC}$, and $\overline{AB}$. These are the second three points of the 9-point circle, and are labeled by G, H, and I in Figure 16.13.

You may notice that the altitudes appear to intersect in a single point (called the orthocenter). You can check that this remains true when you deform the triangle. Before we create the final three points of the 9-point circle, we need a point placed on the orthocenter. As you click to define the point, Sketchpad will only place it as an intersection of two of the lines, but this is fine. Choose two of the altitudes, and mark the orthocenter as J.

The final three points of the circle are the midpoints of the segments $\overline{AJ}$, $\overline{BJ}$, and $\overline{CJ}$. Before we can construct the midpoints, however, we need to add those segments to the diagram, because Sketchpad will only construct the midpoint of an existing segment, not the midpoint between two points. At the moment, we have only the (infinite) altitude lines, and we cannot use them to make a midpoint. So, use the **Segment Straightedge** tool to add a line segment from each vertex to the orthocenter.

Note. You may not be able to see a line segment drawn on top of each altitude line. You can make the segments stand out by making the altitude lines dashed (or another color).

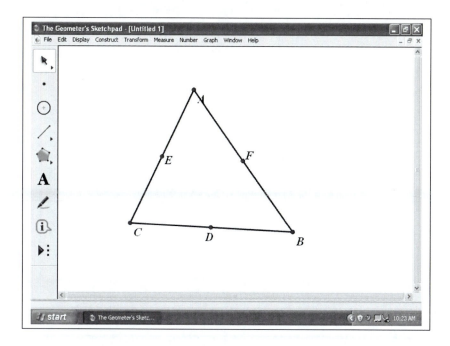

FIGURE 16.12: Geometer's Sketchpad, first three points of the 9-point circle.

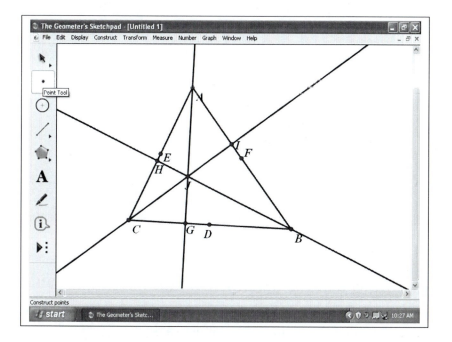

FIGURE 16.13: Geometer's Sketchpad, six points and an orthocenter.

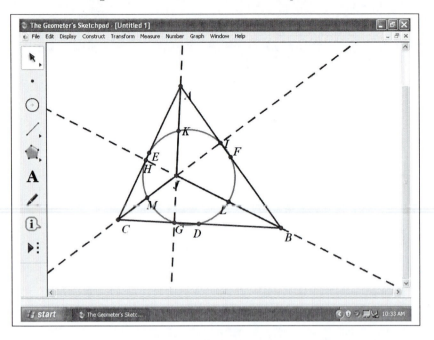

FIGURE 16.14: Geometer's Sketchpad, the 9-point circle.

Right-click each altitude line and choose Dashed from the pop-up menu. Then the still-solid line segments will become apparent.

Using the Arrow tool, select the segment $\overline{AJ}$ and choose Construct/Midpoint. Also create (in a similar way) the midpoint of $\overline{BJ}$ and of $\overline{CJ}$. At this stage, the 9-point circle should be essentially obvious to you.

We can make the circle explicit using the Arc Through 3 Points option from the Construct menu. Use the Arrow tool to select precisely three of the points, say D, M, and E, then choose Construct/Arc Through 3 Points. Use three more points, say E, F, and D, to construct an arc that finishes the circle. See Figure 16.14.

To get something of a hint about what makes this construction work, draw three more segments: $\overline{KD}$, $\overline{LE}$, and $\overline{MF}$. You will notice that these segments are diameters that intersect at the center of the circle. You may wish to change the color or style of these diameters to help them stand out.

Our drawing now has so many parts that you may find that all of the labels create a lot of clutter. Labels can be turned on and off both individually and collectively. To hide all labels, begin with Edit/Select All to select everything in the drawing. Follow with Display/Hide Labels to suppress the labels. If you decide to have labels again later, Display/Show Labels will toggle them back on.

Exporting a drawing to an interactive Web page

An interactive Web page is an especially nice way to share a dynamic drawing, because it lets your audience manipulate a drawing without having to have their own copy of Geometer's Sketchpad. Anyone with a Java-enabled browser can access a drawing and play with it.

To save your drawing as a Web page, choose File/Save As. In the Save As dialog, select Save as type. Change it to HTML/JavaSketchpad Document (*.htm), and save. Sketchpad

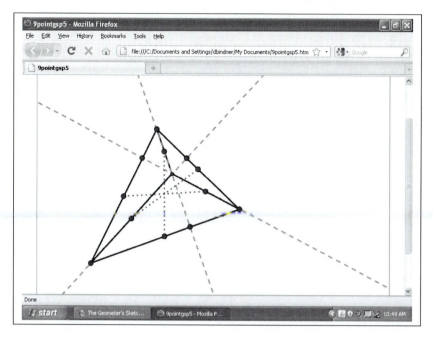

FIGURE 16.15: Geometer's Sketchpad, a JavaSketchpad document.

will create an HTML file that contains instructions for constructing your drawing. It will also create a file called `jsp5.jar` that contains Java code to interpret your dynamic drawing in the Web browser.

Copy your HTML file and the associated `jsp5.jar` file to a Web server to make your drawing accessible as a Web page. Compare with Figure 16.15, which shows the 9-point circle in Firefox.

Note. Some construction elements do not export correctly to JavaSketchpad documents. One of these is the Arc Through 3 Points construction. Notice that the circle is missing from our drawing in Figure 16.15. If we want to see the circle on a Web page, we need to construct it in a different way. One way would be to use the Circle by Center+Point construction, after we have identified the center of the 9-point circle at the intersection of its diameters.

16.5 How to do more elaborate sketches in GeoGebra

The 9-point circle

The 9-point circle makes a nice construction in GeoGebra, just as it does in Geometer's Sketchpad.

To begin the construction, use the New Point tool to create three vertices A, B, and C for a triangle. Follow with the Segment tool to create the sides. (Here's a hint. If you connect B to C first, C to A second, and A to B last, then side a will be across from vertex A and so on according to the usual labeling for triangles.) Use the Midpoint tool to mark

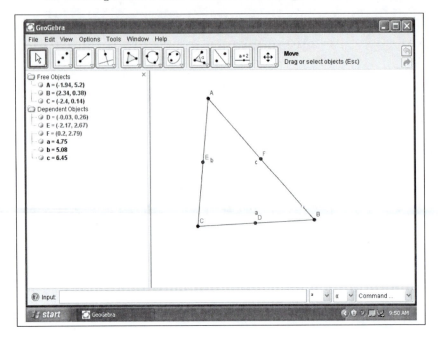

FIGURE 16.16: GeoGebra, first three points of the 9-point circle.

the midpoints, D, E, and F of the sides a, b, and c of the triangle. These are the first three points of the 9-point circle. Your drawing may look something like Figure 16.16.

To find the next three points of the 9-point circle, we need to create the altitudes of the triangle. Each altitude is the line going through a vertex of the triangle and perpendicular to the opposite side. We will use the **Perpendicular Line** tool (fourth button on the toolbar) to create all three. With the **Perpendicular Line** tool selected, click point A and segment $\overline{BC}$ to create the first altitude. The other two altitudes are formed similarly.

We need the points of intersection, so switch to the **Intersect Two Objects** tool (second button on the toolbar). Add a point where the altitudes intersect $\overline{BC}$, $\overline{AC}$, and $\overline{AB}$. These are the second set of three points of the 9-point circle, and are labeled by G, H, and I in Figure 16.17.

You may notice that the altitudes appear to intersect in a single point (called the orthocenter). You can check that this remains true when you deform the triangle. Before we create the final three points of the 9-point circle, we need a point placed on the orthocenter. As you click to define the point, GeoGebra will only place it as an intersection of two of the lines, and it will prompt you for the lines you want to use. Choose two of the altitudes, and mark the orthocenter.

The final three points of the circle are the midpoints of the segments $\overline{AJ}$, $\overline{BJ}$, and $\overline{CJ}$. Unlike Sketchpad, GeoGebra does not require the segments to be explicitly constructed first. The **Midpoint** tool will happily mark the midpoint of two points. So choose the **Midpoint** tool (second button on the toolbar), and click A and J to mark the midpoint of segment $\overline{AJ}$. Mark the other two midpoints in the same way. These form the final points of the 9-point circle. At this stage, the 9-point circle should be essentially obvious to you.

We can make the circle explicit using the **Circle Through Three Points** tool (the sixth button on the toolbar). Click any three of the points, say D, E, and F, and the 9-point circle will appear. See Figure 16.18.

To get something of a hint about what makes this construction work, draw three more

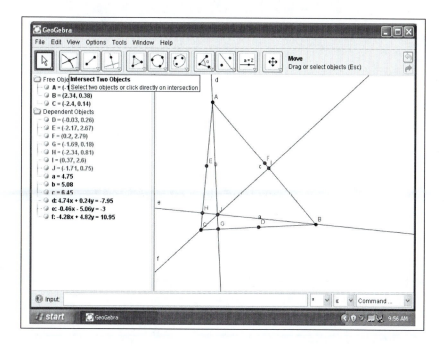

FIGURE 16.17: GeoGebra, second three points of the 9-point circle.

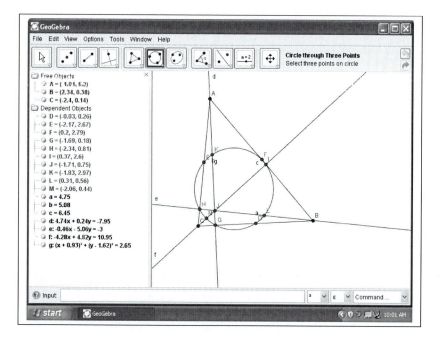

FIGURE 16.18: GeoGebra, the 9-point circle.

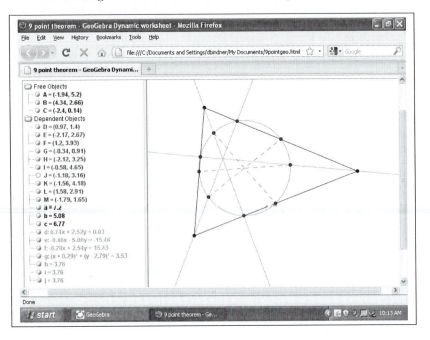

FIGURE 16.19: GeoGebra, a dynamic worksheet in a Web page.

segments: $\overline{KD}$, $\overline{LE}$, and $\overline{MF}$. You will notice that these segments are diameters that intersect at the center of the circle. You may wish to change the color or style of these diameters to help them stand out.

Our drawing now has so many parts that you may find that all of the labels create a lot of clutter. Labels can be turned on and off both individually and in collectively. To hide all labels, begin with Edit/Select All to select everything in the drawing. Follow with Edit/Properties, choose the Basic tab, and deselect the checkbox that says Show Label.

Exporting a drawing to an interactive Web page

An interactive Web page is an especially nice way to share dynamic drawings, because it lets your audience manipulate a drawing without having to have their own copy of Geo-Gebra. Anyone with a Java-enabled browser can access a drawing and play with it.

To save your drawing as a Web page, choose File/Export/Dynamic Worksheet as Webpage (html). After prompting you for some information about the Web page, GeoGebra will create an HTML file and an associated .ggb file (GeoGebra's native file format) that contains instructions for constructing your drawing. It will also create two "jar" files, geogebra.jar and geogebra_properties.jar, that contain the Java code GeoGebra itself.

Copy your HTML file and the associated files to a Web server to make your drawing accessible as a Web page. Compare with Figure 16.19, which shows the 9-point circle in Firefox.

16.6   How to export images from Geometer's Sketchpad and Geo-Gebra

Geometer's Sketchpad and GeoGebra can also be used to create beautiful images for documents. Sketchpad can export images in enhanced metafile format (.emf) that are compatible with programs like Microsoft Word and Microsoft Paint. Choose File/Save As and select Enhanced Metafile in the Save As dialog window.

GeoGebra can save images in EMF format as well. It can also save to Portable Network Graphics (PNG), Portable Document Format (PDF), Encapsulated PostScript (EPS), and Scalable Vector Graphics (SVG). Choose File/Export/Graphics View as Picture and select a format in the Export dialog window.

A color illustration of the 9-point circle construction, as generated by GeoGebra, is shown in Figure 12 of the color insert.

16.7   How to learn more

At dynamicgeometry.com, Key Curriculum Press has an extensive collection of resources, including software for downloading, videos, tutorials, and activities for Geometer's Sketchpad. The welcome videos starring Sketchpad's designer Nick Jackiw are a quick way to see the potential of dynamic geometry software. See:

learningcenter.dynamicgeometry.com

The main Web site of the GeoGebra project is www.geogebra.org. The site contains copies of the program for downloading, PDF documentation, and links to YouTube videos. There are also forums (in multiple languages) where you can ask other GeoGebra users for help, and a Wiki where you can find and share lessons based on GeoGebra.

Exercises

1. Figure out how to get the computer to measure the area of the (outer) quadrilateral and the (inner) parallelogram of Varignon's Theorem to verify that the parallelogram has precisely half the area of the quadrilateral.

2. Let P be a point on the circumcircle of triangle ABC. Let P_1, P_2, and P_3 be the reflections over the sides $\overline{AB}$, $\overline{BC}$, and $\overline{AC}$. Then P_1, P_2, and P_3 all lie on a line l that also passes through the orthocenter of $\triangle ABC$. Create a drawing to demonstrate this.

   Hints: The Transform menu of Geometer's Sketchpad contains a Reflect entry. Use Transform/Mark Mirror first to select the appropriate reflection line. The eighth button on the GeoGebra toolbar has a Mirror object at line tool—select the point first and the reflection line second.

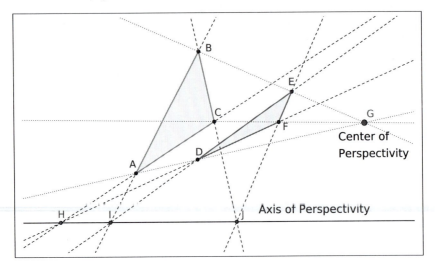

FIGURE 16.20: Desargues' Theorem.

3. Desargues' Theorem, as it applies to the triangles in Figure 16.20, states that the lines $\overleftrightarrow{BE}$, $\overleftrightarrow{CF}$, and $\overleftrightarrow{AD}$ intersect in a common point G if and only if the points H, I, and J lie on a common line. Create a drawing to demonstrate this.

4. Both Geometer's Sketchpad and GeoGebra allow you to create new tools. Construct a custom **Square** tool, and use it to create a diagram similar to Figure 16.21.

   Or, for a more elaborate exercise, search the Internet for A. J. W. Duijvestijn's smallest perfect "simple squared square," a tiling of the 112×112 square using 21 smaller squares of different sizes. Create a drawing of Duijvestijn's tiling.

5. A golden rectangle is a rectangle whose sides are in the proportion $\dfrac{1+\sqrt{5}}{2} : 1$.

   (a) Construct a golden rectangle. Hint: In a unit square, the distance from the midpoint of one side to a far corner is $\dfrac{\sqrt{5}}{2}$.

   (b) Create a custom **Golden Rectangle** tool, and use it to draw a golden spiral of rectangles.

6. Look up Monge's Theorem and create a drawing to demonstrate it. Hint: You may need to hide some objects that are necessary to create the construction but that you don't want to see in your final drawing.

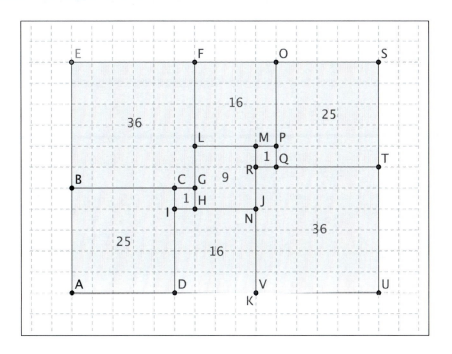

FIGURE 16.21: A rectangle tiled by squares.

Chapter 17

Getting Started with PostScript®

    The purpose of this chapter is to help you begin using PostScript (PS), a page description language with which you can create beautiful, precise graphics. Although PostScript can produce entire pages, interweaving text and pictures, we focus on Encapsulated PostScript (EPS), which is used especially for producing figures for inclusion in other documents (e.g., in LaTeX documents). We present several examples for you to try. To go further with PostScript, you may want to consult the references in the "How to Learn More" section at the end.

17.1   What is PostScript?

    PostScript (PS), created in 1985 by Adobe Systems, Inc., is a printing and imaging system with which you can produce high-quality, beautiful graphics. PostScript is the default graphics language for Unix and GNU/Linux operating systems. PostScript is an interpreted, vector-based graphics language, with many small commands. PostScript is also a programming language, containing the usual programming constructs such as procedures, arrays, and loops. PostScript files can be produced with any text editor (as .eps files), e.g., Notepad. Output can be viewed with a viewer such as the one in PCTeX or a free viewer such as GSview.

The name PostScript is both a play on words (postscript of a letter) and an indication of the postfix order of operations in PostScript. The postfix order of operations derives from the main data structure in the PostScript language: the stack. To understand PostScript, you must understand the stack.

17.2   How to use the stack

The essential data structure in PostScript is the stack, which we can think of like a stack of books. The last item placed on the stack is the first item removed. The stack contains operands, which operators work on. The stack concept ties in nicely with the convention of postfix notation.

Suppose that we list the following objects: 6, the string `cat`, and -100.

```
6 (cat) -100
```

The PostScript interpreter stores the objects on the stack as pictured here.

-100
(cat)
6

Notice that the objects, read from left to right, appear on the stack from bottom to top.

Three useful operators

The `pop` operator removes the top object from the stack. The `exch` operator interchanges the top two objects of the stack. The `dup` operator duplicates the top object of the stack.

For example, starting with the stack from before, we remove the top object.

```
pop
```

Now the stack looks like this.

(cat)
6

Next, we interchange the top two objects of the stack.

`exch`

6
(cat)

Finally, we duplicate the top object of the stack.

`dup`

And now the stack looks like this.

6
6
(cat)

Arithmetic and other basic operations

The arithmetic operations addition, subtraction, multiplication, and division are performed via the operators `add`, `sub`, `mul`, and `div`. For example, let's add two numbers, 6 and 7.

`6 7 add`

The 6 is placed on the stack, followed by the 7. The operator `add` takes the top two numbers off the stack (6 and 7) and places their sum (13) on the stack.

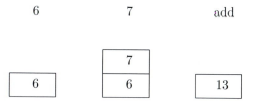

Other mathematical functions are calculated in a similar manner. For example,

`45 sin`

yields the sine of 45 degrees.

17.3 How to make simple pictures

In this section and the following sections, we show several examples of PostScript figures. With these examples, we hope that you will gain an understanding of what can be achieved with PostScript and see some concrete ideas to try out.

FIGURE 17.1: A square.

Coordinates

PostScript pictures are based on Cartesian coordinate geometry. The origin is at the lower-left corner; x increases as you move to the right and y increases as you move up. One unit equals one "big point"; that is, 72 units equal 1 inch.

We begin by describing how to draw lines. To draw a line, begin a new drawing path with `newpath`. Then move the drawing instrument to a point with `moveto`. Then create a line with `lineto`. Finally, after the path has been defined, fill the path with ink by `stroke`.

Example 17.1. We use basic commands to draw a square of side length 1 inch.

```
%!PS-Adobe-3.0 EPSF-3.0
%%Title: Square
%%BoundingBox: 0 0 144 144

% draw a 1 inch square
newpath                      % start a new path (clears any old paths)
36 36 moveto                 % beginning point of the path at (36,36)
108 36 lineto                % bottom of square
108 108 lineto               % right side
36 108 lineto                % top side
36 36 lineto                 % left side
stroke                       % trace the path (with black ink)
```

The picture is shown in Figure 17.1.

The % (percentage) sign is used for comments. Certain comments give structural information about a file. In our input above, the first three lines are an identification, a title, and a bounding box. The bounding box gives coordinates for the lower-left corner and upper-right corner of the picture.

There is some flexibility in writing expressions. For example, we can draw the above square with *relative* `moveto` and `lineto` commands called `rmoveto` and `rlineto`, respectively. Also, we can close the path with a `closepath` command, rather than explicitly making a line back to the starting point.

```
newpath                      % start a new path
```

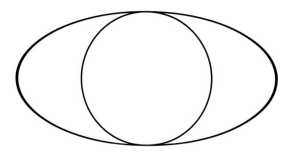

FIGURE 17.2: A circle and an ellipse.

```
36 36 moveto             % beginning point of the path at (36,36)
72 0 rlineto             % extend path 1 inch right
0 72 rlineto             % 1 inch up
-72 0 rlineto            % 1 inch left
closepath                % back to starting point
stroke                   % trace the path (with black ink)
```

Simple transformations

It's sometimes convenient to make transformations of the coordinate system. The basic types of transformations are translations, scalings, and rotations.

Example 17.2. We illustrate translation and scaling with a drawing of a circle and an ellipse, as shown in Figure 17.2.

```
%!PS-Adobe 3.0 EPSF-3.0
%%Title: Circle and ellipse
%%BoundingBox: 0 0 216 216

% move coordinate system over and up 1.5 inches
108 108 translate

% draw circle at (0,0) with 0.75 inch radius (54 points)
newpath 0 0 54 0 360 arc stroke

% stretch coordinate system in x by factor of 2
2 1 scale

% draw ellipse (a circle stretched by the scale)
newpath 0 0 54 0 360 arc stroke
```

The command 108 108 translate moves the coordinate system so that 108 108 is the new origin.

The operator arc takes five arguments: the two coordinates of the center of the arc, the radius of the arc, and the beginning and ending reference angles of the arc. In the previous

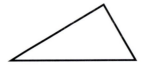

FIGURE 17.3: A rotated 30°–60°–90° triangle.

code, both arcs have center $(0,0)$, but actually appear at $(108, 108)$ because of the prior transformation.

The command 2 1 scale scales the x-axis by a factor of 2 and leaves the y-axis unchanged. Drawing a "circle" now produces an ellipse.

Note. You can use the concept of scaling an axis to solve a math problem. Show how to inscribe in a given ellipse a triangle of maximum area. Can you draw such a triangle on the diagram in Figure 17.2?

The command **rotate** rotates the coordinate system a given number of degrees in the counterclockwise direction.

Example 17.3. We draw a 30°–60°–90° triangle in a coordinate system rotated 30° counterclockwise about the origin.

```
%!PS-Adobe-3.0 EPSF-3.0
%%Title: 30-60-90 triangle rotated
%%BoundingBox: 0 0 144 144

% move origin 1 inch right and up
72 72 translate

% rotate coordinate system 30 degrees counterclockwise
30 rotate

% draw triangle
newpath
0 0 moveto               % 1st vertex of triangle
36 3 sqrt mul  0 rlineto % extend path 36*sqrt(3) to the right
0 -36 rlineto            % extend path 36 points down (1/2 inch)
closepath                % back to 1st vertex
stroke                   % trace path (with black ink)
```

The result is shown in Figure 17.3.

Procedures and variables

A procedure is defined by a command such as

/p {*operations*} def

(this command defines a procedure called p that performs *operations*). Procedures can take arguments from the stack.

A variable is similar to a procedure but without the {} delimiters, and it holds a value rather than commands.

Example 17.4. We make a series of shaded squares using a procedure called inch and a procedure called square. The inch procedure takes a number off the stack and multiplies it by 72, effectively converting values from points to inches. The square procedure takes a number off the stack and sets sidelength (a variable) equal to that number. It does this by exchanging the number with the variable name (via the exch operator) and then making the definition.

The squares are filled in with the fill command. In the black-and-white world, setgray defines a gray level from 0 to 1, where 0=black and 1=white. Notice that each level of gray is opaque, completely covering any graphical elements beneath it. Thus, in any picture, an "erasing" effect can be achieved by drawing over dark images with white (1 setgray).

```
%!PS-Adobe-3.0 EPSF-3.0
%%Title: Shaded squares
%%BoundingBox: 0 0 144 144


% inch procedure --- multiplies the argument by 72
/inch {72 mul} def

% makes a square path
/square {
  /sidelength exch def      % put argument in variable sidelength
  sidelength 0 rlineto      % extend path right
  0 sidelength rlineto      % up
  sidelength neg 0 rlineto  % left
  0 sidelength neg rlineto  % down
} def

% draw first square
newpath
0.5 inch 0.5 inch moveto    % in points: (36,36)
1 inch square               % call procedure with 72
0 setgray                   % set color to black (0% white)
fill                        % fill in the square

% draw second square
newpath
0.75 inch 0.5 inch moveto   % in points: (54,36)
0.75 inch square            % call procedure with 54
0.25 setgray                % set to dark gray (25% white)
fill                        % fill in the square
```

FIGURE 17.4: Shaded squares.

```
% draw third square
newpath
1 inch 0.5 inch moveto        % in points: (72, 36)
0.5 inch square               % call procedure with 36
0.5 setgray                   % set to medium gray (50% white)
fill                          % fill in the square

% draw fourth square
newpath
1.25 inch 0.5 inch moveto     % in points: (90,36)
0.25 inch square              % call procedure with 18
0.75 setgray                  % set to light gray (75% white)
fill                          % fill in the square
```

The result is shown in Figure 17.4.

Clipping

Clipping defines a closed path that serves as a "universe" for all further graphics insertions. A clipping path is like laying a window on the picture. With a clipping path in effect, we can draw anywhere in the picture, but only the parts in the window will show.

Example 17.5. We create a Venn diagram using a clipping path.

```
%!PS-Adobe-3.0 EPSF-3.0
%%Title: Venn diagram
%%BoundingBox: 0 0 144 144

% define left, right, and bottom circle procedures
/leftcircle {72 90 36 0 360 arc} def
/rightcircle {108 90 36 0 360 arc} def
/bottomcircle {90 54 36 0 360 arc} def
```

```
% draw shaded circles 85% white
0.85 setgray
newpath leftcircle fill
newpath rightcircle fill
newpath bottomcircle fill

% draw shaded two-circle intersections 70% white
0.70 setgray
gsave
  % make clipping path of left circle
  newpath leftcircle clip
  % draw shaded intersection of left and right circles
  newpath rightcircle fill
  % draw shaded intersection of left and bottom circles
  newpath bottomcircle fill
grestore
gsave
  % make clipping path of right circle
  newpath rightcircle clip
  % draw shaded intersection of right and bottom circles
  newpath bottomcircle fill
grestore

% draw shaded three-circle intersection 55% white
0.55 setgray
gsave
  % make clipping path of intersection of left and right circles
  newpath leftcircle clip
  newpath rightcircle clip
  % draw shaded intersection of all circles
  newpath bottomcircle fill
grestore

% draw circumferences of circles black (0% white)
0 setgray
2 setlinewidth
newpath leftcircle stroke
newpath rightcircle stroke
newpath bottomcircle stroke
```

The diagram is shown in Figure 17.5.

Note. The operator **gsave** saves the current graphics state (including current path, current point, clipping path, font, and coordinate system), and **grestore** restores the graphics state to the last saved state.

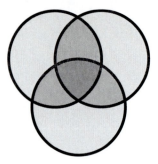

FIGURE 17.5: Venn diagram.

17.4   How to add text to pictures

Text is inserted into a picture by applying the `show` operator to a text string. But first, a font must be chosen using `findfont`, `scalefont`, and `setfont` operators. For example, the command

```
/Courier findfont 12 scalefont setfont
```

chooses a Courier font of size 12 points. The `findfont` operator puts the Courier font on the stack. This is followed on the stack by 12. The `scalefont` operator takes the font and the 12 and replaces them on the stack with a scaled font. Finally, the scaled font is made the current font by the `setfont` operator.

Example 17.6. We put labels on the square of Example 17.1. The result is shown in Figure 17.6.

```
%!PS-Adobe-3.0 EPSF-3.0
%%Title: Square with labels
%%BoundingBox: 0 0 144 144

72 72 translate

% draw square
newpath
-36 -36 moveto
36 -36 lineto
36 36 lineto
-36 36 lineto
-36 -36 lineto
stroke

% make labels
/Courier findfont 12 scalefont setfont
-3 27 moveto (N) show
```

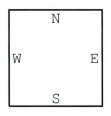

FIGURE 17.6: A labeled square.

```
-3 -33 moveto (S) show
27 -3 moveto (E) show
-33 -3 moveto (W) show
```

Note. Other commonly used fonts are Helvetica, Times-Roman, Times-Italic, and Symbol.

17.5 How to use programming constructs

PostScript supports the usual programming constructs such as loops, arrays, and conditional branching.

Loops

PostScript allows for loops via a **repeat** construction or a **for** construction.

n {*procedure*} **repeat**

The procedure is repeated *n* times.

min increment max {*procedure*} **for**

Successive values are placed on the stack starting with *min*, incrementing by *increment*, until reaching *max*. A **for** loop leaves the counter on the stack, to be used by the procedure.

Example 17.7. Here is a representation of a hypercube obtained by drawing a square, then rotating the coordinate system 45° counterclockwise, and repeating for a total of eight times. The result is shown in Figure 17.7.

```
%!PS-Adobe-3.0 EPSF-3.0
%%Title: Hypercube
%%BoundingBox: 0 0 108 108
```

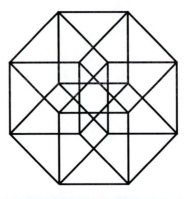

FIGURE 17.7: A hypercube.

```
% move origin over and up 0.75 inches
54 54 translate

% square procedure 40x40 points
/square {
  40 0 rlineto
  0 40 rlineto
  -40 0 rlineto
  0 -40 rlineto
} def

% draw figure
8 {                                     % number of times to loop
  newpath                               % clear old paths
  2 sqrt 1 sub 20 mul -20 moveto        % start at 20(sqrt(2)-1), -20
  square                                % create path for square
  stroke                                % trace it (in black ink)
  45 rotate                             % rotate coordinates for next
} repeat                                % time through the loop
```

Example 17.8. We use a double loop that draws a picture of a triangular lattice. The counters left on the stack by the for loops are set as the values of i and j.

```
%!PS-Adobe-3.0 EPSF-3.0
%%Title: Triangular lattice
%%BoundingBox: 0 0 90 90

% zoom x,y by factor of 5
5 5 scale

% draw a triangle of dots
0 1 9 {             % count through 0,1,...,9  (rows)
  /j exch def       % put counter in variable j
```

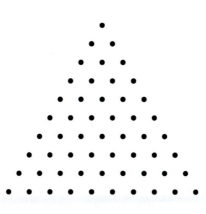

FIGURE 17.8: A triangular lattice.

```
 0 1 9 j sub {      % count through 0,1,...,(9-j)  (columns)
    /i exch def     % put counter in variable i

    % put a filled circle at coordinates (2i+j, sqrt(3)*j)
    newpath  2 i mul j add  3 sqrt j mul  0.25 0 360 arc fill
 } for
} for
```

The picture is shown in Figure 17.8.

Example 17.9. We employ a loop to graph a function by "plotting points."

```
%!PS-Adobe-3.0 EPSF-3.0
%%Title: Function plot
%%BoundingBox: 0 0 350 150

% define function, interval endpoints, etc.
/function {180 mul 3.14 div sin} def   % postfix for sin(x*180/3.14)
/leftendpoint 0 def
/rightendpoint 6.28 def
/magnification 50 def
/increment 0.01 def

% make transformations
magnification dup scale                % scales x and y by 50
1 magnification div setlinewidth       % line width of 1/50 = 1 point
0.5 1.5 translate

% draw coordinate axes
newpath
-0.5 0 moveto 6.5 0 lineto             % x-axis
0 -1.5 moveto 0 1.5 lineto             % y-axis
stroke
```

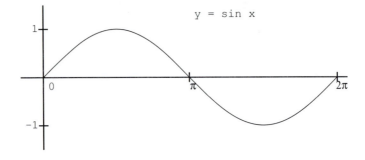

FIGURE 17.9: A sine curve.

```
% draw ticks on axes
newpath
-0.1 -1 moveto 0.1 -1 lineto                  % at y=-1
-0.1 1 moveto 0.1 1 lineto                    % at y=1
3.14 -0.1 moveto 3.14 0.1 lineto              % at x=3.14
3.14 2 mul -0.1 moveto 3.14 2 mul 0.1 lineto  % at x=6.28
stroke

% make labels
/Courier findfont 12 magnification div scalefont setfont
-0.25 0.95 moveto (1) show
-0.40 -1.05 moveto (-1) show
0.1 -0.25 moveto (0) show
3.25 1.25 moveto (y = sin x) show
/Symbol findfont 12 magnification div scalefont setfont
3.15 -0.25 moveto (\160) show
6.25 -0.25 moveto (2\160) show

% This makes a single point from x,y on the stack
/point {increment 0 360 arc} def

leftendpoint increment rightendpoint { % count from 0 to 6.28 by .01
   dup                % duplicate the counter value on stack
   function           % call function on 2nd value (to get y)
   newpath            % clear old paths
   point              % make point at (x,y)
   fill               % and fill point (with black ink)
} for
```

The graph is shown in Figure 17.9.

You can create similar programs for plotting parametric functions and functions in polar coordinates. For example, the commands

```
0 0.5 360 {           % count from 0 to 360 by 0.5
   newpath            % clear old paths
```

```
% remember: the loop counter is still on the stack
sin 1 add 72 mul 0 0.5 0 360 arc % dot at ((sin(t)+1)*72, 0)
fill                            % with radius 0.5 filled

  0.5 rotate        % rotate 0.5 degrees each time
} for
```

draw the cardioid $r = 1 + \sin\theta$, for $0 \leq \theta < 360°$, in polar coordinates.

Arrays

PostScript allows for the creation and manipulation of arrays. An array of size n is indexed as [0 1 2 ... n-1]. An array's values are assigned and retrieved with the put and get operators, respectively. An array's entries can be any objects, e.g., numbers, strings, even other arrays.

For example, to put the value 99 in position 50 in the array birds, use the commands:

```
birds 50 99 put
```

To get the object in position 50 in the array birds, use:

```
birds 50 get
```

Conditional branching

PostScript allows for conditional branching via an if construction or an ifelse construction.

Example 17.10. We make a graphical representation of the Sieve of Eratosthenes, using an array and conditional branching. The picture shows prime numbers as white boxes and composite numbers (proper multiples of the primes 2, 3, 5, and 7) cancelled out in gray.

```
%!PS-Adobe-3.0 EPSF-3.0
%%Title: Sieve
%%BoundingBox: 0 0 234 234

20 20 translate

% square procedure
/square {
  20 0 rlineto
  0 20 rlineto
  -20 0 rlineto
  0 -20 rlineto
} def

% create an array with entries numbered 0,...,100
/numbers 101 array def

% initialize array --- each entry starts out with 1
2 1 100 {              % count through 2,3,...,100
 numbers exch          % exchange array with count on stack
 1 put                 % and put 1 into that array location
} for
```

```
% run sieve --- When finished, entries in the array
% that still hold 1 are primes.  Other entries hold 0.5.
2 1 10 {                 % count through 2,3,...,10
  /i exch def            % put counter in variable i

  numbers i get 1 eq {   % if numbers[i]==1 then
    i 2 mul i 100 {      % count through 2i,3i,...,100
      numbers exch       % exchange array with counter on stack
      0.5 put            % and put 0.5 in that location
    } for
  } if
} for


% draw grid
2 1 100 {                % count though 2,3,...,100
  /i exch def            % put counter in variable i
  newpath                % erase old paths

  % move to (((i-1) mod 10)*20, 180-((i-1) idiv 10)*20)
  i 1 sub 10 mod 20 mul  180 i 1 sub 10 idiv 20 mul sub
  moveto
  square                 % make path for square

  gsave
    numbers i get        % get the ith entry in array
    setgray              % and use it to set gray level
    fill                 % fill the square (gray or white)
  grestore

  0 setgray              % set color black
  stroke                 % trace the square
} for
```

The result is shown in Figure 17.10.

17.6   How to add color to pictures

Color is defined with the setrgbcolor operator. This operator takes a triple of numbers that denote an additive mixture of red, green, and blue color light. The numbers are between 0 and 1, with 0 denoting no light of the given color and 1 denoting the maximum amount of light. For instance, 0 0 1 setrgbcolor corresponds to pure blue light.

Example 17.11. We make a color wheel.

```
%!PS-Adobe-3.0 EPSF-3.0
%%Title: Color wheel
%%BoundingBox: 0 0 270 270

135 135 translate        % move origin up and over
```

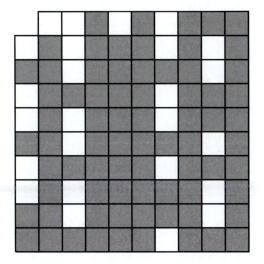

FIGURE 17.10: Sieve of Eratosthenes.

```
30.0 setlinewidth          % 30pt wide lines (fat lines)

% draw green--red third of wheel
0 0.25 0.75 {              % count through 0,0.25,0.5,0.75
  /lambda exch def
  lambda 1 lambda sub 0 setrgbcolor % (lambda,1-lambda,0)
  newpath 90 0 moveto 108 0 lineto stroke
  30 rotate                % rotate 30 degrees for next loop
} for

% draw red--blue third of wheel
0 0.25 0.75 {              % count through 0,0.25,0.5,0.75
  /lambda exch def
  1 lambda sub 0 lambda setrgbcolor % (1-lambda,0,lambda)
  newpath 90 0 moveto 108 0 lineto stroke
  30 rotate                % rotate 30 degrees for next loop
} for

% draw blue--green third of wheel
0 0.25 0.75 {              % count through 0,0.25,0.5,0.75
  /lambda exch def
  0 lambda 1 lambda sub setrgbcolor % (0,lambda,1-lambda)
  newpath 90 0 moveto 108 0 lineto stroke
  30 rotate                % rotate 30 degrees for next loop
} for

% make labels
/Courier findfont 12 scalefont setfont
```

```
0 setgray                               % set color black
30 rotate -10 120 moveto (red) show
-120 rotate -15 120 moveto (green) show
-120 rotate -10 120 moveto (blue) show
```

See Figure 13 in the color insert.

Note. A color wheel is but a one-dimensional slice of the three-dimensional space of color combinations in the RGB model.

Example 17.12. We picture seven rings of three different colors, with no two rings linked but all seven impossible to unlink. Notice that three small corrections near the end of the program make the rings pass over and under each other properly.

```
%!PS-Adobe-3.0 EPSF-3.0
%%Title: Seven interlocking rings
%%BoundingBox: 0 0 466 466


233 233 translate       % move origin over up 3.2 inches
15.0 setlinewidth       % lines are about 0.2 inches wide


% draw middle ring
0 0.25 0 setrgbcolor    % dark green
newpath
0 0 108 0 360 arc       % circle with radius 1.5 inches
stroke


% draw three outer rings
0.95 0.95 0.40 setrgbcolor % light yellow
3 {                     % number of times to loop
  newpath
  108 0 108 0 360 arc   % circle with radius 1.5 inches
  stroke
  120 rotate            % rotate coordinates for next loop
} repeat


60 rotate
% draw three more outer rings
1 0 0 setrgbcolor       % red
3 {                     % number of times to loop
  newpath
  108 0 108 0 360 arc   % circle with radius 1.5 inches
  stroke
  120 rotate            % rotate coordinates for next loop
} repeat


% make three corrections to middle ring (to bring it over
% the red rings, but not over the yellow rings)
-60 rotate
0 0.25 0 setrgbcolor    % dark green
3 {                     % number of times to loop
  newpath
  0 0 108 -10 10 arc    % only from angle -10 to 10
```

```
  stroke
  120 rotate              % rotate coordinates for next loop
} repeat
```

The resulting image ("Seven interlocking rings") is displayed on the front cover of this book.

17.7   More examples

We give a few more examples of PostScript figures.

Combining what we know

Example 17.13. We draw a two-colored complete graph on 17 vertices, using an array, loops, branching, and color. The coloring has the property that there exist no four vertices whose six edge connections are all the same color.

```
%!PS-Adobe-3.0 EPSF-3.0
%%Title: Complete graph of order 17 with colored edges
%%BoundingBox: 0 0 180 180

% define constants and array
/radius 81 def
/angle 360 17 div def
/vertex 17 array def

% make transformations
90 90 translate
0.5 setlinewidth

% Define the array of vertices.  Each entry of the
% array will be an array [x,y] of the coordinates of
% that vertex of the graph.
0 1 16 {          % count through 0,1,...,16
  /i exch def     % put counter in variable i

  % puts [cos(i*angle)*radius, sin(i*angle)*radius] into entry i
  vertex i
  [i angle mul cos radius mul  i angle mul sin radius mul]
  put
} for
```

```
% draw vertices
0 1 16 {             % count through 0,1,...,16
  /i exch def        % put counter in variable i
  newpath            % clear old paths

  vertex i get       % get [x,y] on stack
  0 get              % get x coordinate on stack

  vertex i get       % get [x,y] on stack
  1 get              % get y coordinate on stack

  2 0 360 arc        % uses x,y on stack with radius 2
  fill               % fill dot with black
} for

% draw edges
0 1 16 {
  /i exch def

  i 1 add 1 16 {
    /j exch def
    /d {j i sub} def

    % Set color to green if the difference between i and j is
    % 1,4,9,16,8,2,15, or 13.  Red for the other differences.
    d 1 eq d 4 eq d 9 eq d 16 eq d 8 eq d 2 eq d 15 eq d 13 eq
    or or or or or or or {
       0 1 0 setrgbcolor          % green
    }{
       1 0 0 setrgbcolor          % red
    } ifelse

    % make line from vertex i to vertex j and stroke it
    newpath
    vertex i get 0 get  vertex i get 1 get  moveto
    vertex j get 0 get  vertex j get 1 get  lineto
    stroke
  } for
} for
```

See Figure 14 in the color insert.

Affine transformations

In PostScript, we can make affine transformations of the plane, including translations, scalings, rotations, and reflections. The common operators for the first three of these transformations are translate, scale, and rotate, while a reflection is a scaling where one of the scale factors is 1 and the other is −1. The general operator is concat, which takes six arguments defining the affine transformation. The command [a b c d e f] concat

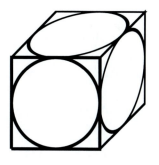

FIGURE 17.11: A cube with circles.

yields the transformation

$$\begin{bmatrix} x & y \end{bmatrix} \mapsto \begin{bmatrix} x & y \end{bmatrix} \begin{bmatrix} a & b \\ c & d \end{bmatrix} + \begin{bmatrix} e & f \end{bmatrix}.$$

Example 17.14. We make a cube with a circle on each visible face. The two uses of concat are transformations called shears.

```
%!PS-Adobe-3.0 EPSF-3.0
%%Title: Cube with circles
%%BoundingBox: 0 0 54 54

% square procedure
/square {
  30 0 rlineto
  0 30 rlineto
  -30 0 rlineto
  0 -30 rlineto
} def

4.5 4.5 translate

% draw cube front side
newpath 0 0 moveto square stroke
newpath 15 15 15 0 360 arc stroke

% draw cube top side
gsave
  [1 0 0.5 0.5 0 30] concat
  newpath 0 0 moveto square stroke
  newpath 15 15 15 0 360 arc stroke
grestore
```

```
% draw cube right side
[0.5 0.5 0 1 30 0] concat
newpath 0 0 moveto square stroke
newpath 15 15 15 0 360 arc stroke
```

The picture is shown in Figure 17.11.

Recursion

PostScript allows procedural recursion; that is, a procedure can call itself.

Example 17.15. We create the Sierpinski triangle (a fractal) using recursion. We start with a black filled triangle, and then fill the medial triangle white. The process is repeated recursively on each of the remaining black triangles. We do the recursion to depth 8.

```
%!PS-Adobe-3.0 EPSF-3.0
%%Title: Sierpinski triangle
%%BoundingBox: 0 0 200 200

% Sierpinski triangle procedure.  Expects coordinates
% for 3 vertices to be on the stack, i.e.,
% x1,y1, x2,y2, and x3,y3, and number of levels
% of recursion to descend, n.
/sierpinskitriangle {
  /n exch def
  /y3 exch def /x3 exch def
  /y2 exch def /x2 exch def
  /y1 exch def /x1 exch def

  % fill triangle black
  0 setgray
  newpath
  x1 y1 moveto
  x2 y2 lineto
  x3 y3 lineto
  closepath fill

  % fill center triangle white
  1 setgray
  newpath
  x1 x2 add 2 div y1 y2 add 2 div moveto
  x2 x3 add 2 div y2 y3 add 2 div lineto
  x3 x1 add 2 div y3 y1 add 2 div lineto
  closepath fill

  % call this function recursively if n>0
  % to do the subtriangles
  n 0 gt {
    x1 y1
    x1 x2 add 2 div y1 y2 add 2 div
    x1 x3 add 2 div y1 y3 add 2 div
    n 1 sub
```

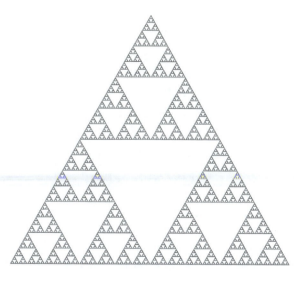

FIGURE 17.12: Sierpinski triangle.

```
    x2 y2
    x1 x2 add 2 div y1 y2 add 2 div
    x2 x3 add 2 div y2 y3 add 2 div
    n 1 sub

    x3 y3
    x1 x3 add 2 div y1 y3 add 2 div
    x2 x3 add 2 div y2 y3 add 2 div
    n 1 sub

    sierpinskitriangle
    sierpinskitriangle
    sierpinskitriangle
  } if
} def

% GO!
0 0 100 173 200 0 8 sierpinskitriangle
```

The picture is shown in Figure 17.12.

17.8　How to learn more

Many aspects of PostScript are not discussed in this introduction, such as word art and dictionaries. Here are some resources for you to investigate to learn more.

The definitive books about PostScript are the "Blue Book" (tutorial) [3], the "Red Book" (reference manual) [2], and the "Green Book" (programming guide) [1]. Some good beginning books (in addition to the "Blue Book") are [52], [45], and [38].

A good way to learn PostScript is to use it to make figures. You might try producing a figure illustrating, for instance, a triangular lattice, the Fano Configuration, the Desargues Configuration, a tiling of Poincaré's hyperbolic plane, or an approximation to a Peano area-filling curve.

For information on the RGB color model (and the related CMYK model), see:

www.colorcube.com

For connections between PostScript and geometry, see "A Manual for Mathematical PostScript" at [7] or:

www.sunsite.ubc.ca/DigitalMathArchive/Graphics/text/www

For inspiration about mathematics and art, [41] is a wonderful book.

Exercises

1. What is left on the stack after the following PostScript input?

```
3 dup div neg 10 add dup div pop 10
```

2. What is wrong with the following input?

```
3 2 add 6 sub 9 add add
```

3. What does this input draw?

```
newpath
10 10 moveto
20 20 moveto
100 20 lineto
100 100 lineto
20 20 lineto
stroke
```

4. Draw a blank Sudoku board, as shown below.

5. Draw a regular heptadecagon (17 sides). Hint: You may want to use rotation and a loop.

6. Draw a circle filled so that its center is black and the gray level increases radially toward its circumference. Hint: Use a loop to draw concentric circles.

7. Draw a *square* lattice of dots, similar to the triangular lattice of dots in Figure 17.8.

8. What does this input draw?

```
newpath
30 30 20 0 360 arc
gsave
   0.5 setgray
   fill
   30 30 10 0 360 arc
   gsave
      1 setgray
      fill
   grcstore
   stroke
grestore
stroke
```

9. Draw a picture of the cardioid $r = 1 + \sin\theta$, for $0 \le \theta < 360°$, in polar coordinates.

10. Use Horner's method to graph the polynomial

$$y = x^6 + 3x^5 - 2x^4 + x^3 + 7x^2 - 3x + 1, \quad -2 \le x \le 2.$$

11. Plot the function $y = \sin x$, for $0 \le x \le \pi$, with the area bounded by the curve and the x-axis shaded gray.

12. Make a picture, as in Figure 15 of the color insert, showing twelve pentominoes of different colors packed into a 10×6 box.

13. Redo Example 10.6, the tessellation of crosses. Hint: If you scale by $\left(\dfrac{72}{2.54}\right) 0.3$, your PostScript program can use the same units as the PSTricks code. Use a loop within a loop, and put the initial points of the crosses at coordinates $(3i - j, 3j + i)$, with $1 \le i \le 4, 1 \le j \le 3$.

14. Make a picture showing a colorful dissection proof of the Pythagorean Theorem, as in Figure 16 of the color insert. Hint: Use a clipping path.

Chapter 18

Getting Started with Computer Programming Languages

In this chapter, we take a look at some popular programming languages and the general issue of programming in mathematics.

18.1 Why program?

In mathematics, many things contribute to new discoveries. Inspiration plays a role, as does luck. But hard work and experimentation are two of the most important factors, and technology can help make mathematical work easier and experimentation faster.

It's no accident that the most popular software applications for mathematics are also fully capable programming languages. In a programming language we can perform elaborate calculations that we could not express with only algebraic formulas. The ability to program also allows us to easily repeat a test or experiment dozens, hundreds, or even thousands of times.

For example, consider a Linear Algebra student experimenting with matrices. Perhaps she's playing with a simple idea: that you always get a symmetric answer when you multiply a matrix by its transpose. She might test the idea by generating a few matrices in MATLAB or Octave.

Example 18.1. Checking a matrix conjecture by hand.

```
1> A = rand(3)
A =

    0.58695    0.74530    0.71926
    0.99214    0.57297    0.81125
    0.18276    0.92548    0.50683

2> A*A'
ans =

    1.4173    1.5929    1.1616
    1.5929    1.9707    1.1228
    1.1616    1.1228    1.1468
```

Two or three experiments like this will probably convince a student of the truth of a simple claim, but the skeptical student might want to try further examples with larger matrices. With just a little bit of programming, she can easily verify ten (or more if she wants) 7×7 examples.

Example 18.2. Here we use a loop to check that ten random examples are equal to their transpose.

```
3> for i = 1:10
>    A = rand(7);                    # make a random 7x7 matrix
>    B = A*A';                       # multiply by transpose
>    X(i) = max( max( abs(B - B') )); # subtract the transpose
>  end                               # and get largest entry

4> X
X =

    0  0  0  0  0  0  0  0  0  0
```

Every time, the difference between the matrix B and its transpose is zero; the matrices are all symmetric.

18.2   How to choose a language

With so many computer languages available, it can be a daunting task to choose a language for your mathematical work, and not every language is convenient for every project. Worse, different experts will disagree on the appropriate language to use for a project because they have different backgrounds or tastes. While any programming language can be forced to accomplish your project goals, at least theoretically, each language has certain tasks that it does best. So it pays to pick a good language.

For simple experiments or projects, it usually makes sense to use the language that is

	Level	Math	Speed	Stand-alone
Computer Algebra System	5	5	2	1
MATLAB/Octave	4	4	3	2
Perl/Python/Ruby	4	3	3	3
Java	3	3	4	4
Fortran/C/C++	2	3	5	5
Assembler	1	1	5	5
JavaScript	3	2	1	3

TABLE 18.1: Relative language level, math content, speed potential, and suitability for stand-alone programs. Highest/best is indicated by 5.

part of your favorite mathematical software. If you use Mathematica for day-to-day work, then the programming language built in to Mathematica probably makes sense for small programs. If you are currently using Maple, Maxima, MATLAB, or Octave for your work, then those are the right languages to think about first for your programs.

Choosing the language you are already using is likely to be smart in part because there is probably a good reason you are working with that platform in the first place. You might be using Octave because your work involves a lot of matrices (as in Example 18.2) and Octave specializes in matrix manipulation. It is precisely this feature that makes Octave a good candidate for your computer programs as well.

There is one other advantage to staying with the software that you already use. Even if you haven't learned to write programs in your favorite mathematics software, you already know some of the syntax (the specific rules and idiosyncrasies of that particular language). If you use Mathematica, for example, you know that lists are denoted by curly braces, whereas Maxima users know to use square brackets for lists.

The larger your program or project becomes, the more likely you are to benefit by switching to a different language to take advantage of specific features of that language. If you are willing to consider new languages, here are some advantages and disadvantages you might think about. For a summary of the pluses and minuses of various languages, see Table 18.1.

Mathematica, Maple, and Maxima

Having a computer algebra system underneath your programming language can be immensely powerful for mathematical programming. Systems like Mathematica, Maple, and Maxima can do things that would have to be programmed by hand in other languages, like evaluation of expressions and symbolic computation (such as derivatives).

You have an extensive library of functions at your disposal when using a computer algebra system. For example, you may want to test a number at some point of your program to see if it is prime, or you may want to compute a determinant. These kinds of computations are likely to be built-in parts of the language when using a computer algebra system as your underlying programming language.

For many projects, computer algebra systems are an obvious first choice for a programming language, but there is at least one way they may be deficient. If you wish your final program to be a "stand-alone" application, then using a computer algebra system may not be the best choice. Mathematica, for example, is famously organized around its idea of the electronic "mathematical notebook," so any programming you do will generally be in the context of mathematical notebooks. If you share your notebooks with others, they will need their own copy of the Mathematica engine (either in the form of Mathematica itself or the free Mathematica Player). Since computer algebra systems are large programs, using them as the foundation for an independent "stand-alone" program can be unwieldy. It may be possible, but that is not how the programs are commonly used.

Note. In the category of computer algebra systems, Maxima is interesting because source code is available. Maxima, which is a language of its own, was itself written in another language called Lisp. Thus, programmers essentially have two possible languages in Maxima. Most programming projects will not require work at the Lisp level, but the ambitious person who wants to customize Maxima extensively can make changes that "feel like the real thing" by writing code in Lisp and incorporating that code directly into Maxima.

MATLAB and Octave

MATLAB and Octave, which implement versions of the same programming language, have some of the same strengths that the full computer algebra systems have because they are also designed for mathematical work. Matrices and matrix operations are core parts of the language, which means that operations like finding the rank of a matrix or solving a linear system do not have to be coded by you.

Octave has the advantage of being free software, and Octave itself is written in C++. So if you wish to modify Octave, you have much the same potential as programmers who might wish to modify Maxima.

As a commercial product, MATLAB is not readily customizable. However, MATLAB has two particular advantages as a programming language. First, the graphical layer is more flexible than the output engine of Octave. If your program implements interactive input and output through a graphical interface, then MATLAB may be a better choice.

The other strong advantage of MATLAB is its collection of "toolboxes" devoted to particular topics. If your project uses wavelets, for example, the Wavelet Toolbox for MATLAB may be very attractive.

Octave and MATLAB are common language choices for programs that compute numerical solutions to problems, incorporate a great deal of matrix work, or involve signal processing. One drawback of the way the languages are implemented (as scripting languages) is that they may not be fast enough for some kinds of work. Still, it often makes sense to create a prototype in MATLAB or Octave first, then later re-implement the program in a compiled language like Fortran or C for optimal efficiency.

R

If your programming project is heavily statistics based, then R is a promising choice. Not only are existing statistical methods likely to be incorporated already into the language, but new research is also often implemented in R. So the language is growing with the field, and you may find other researchers (and potential collaborators) are already using R for similar work.

Perl, Python, and Ruby

Perl, Python, and Ruby are sometimes described as high-level or very high-level languages. Unlike the languages discussed up to this point, these three are not dedicated to mathematical work but are instead general-purpose programming languages. As higher-level languages, Perl, Python, and Ruby are generally more convenient to program in than mid-level languages like C. More things are handled by a higher-level language and do not have to be coded by the programmer.

Although Perl, Python, and Ruby do not focus on mathematical programming, each of these languages comes with a huge collection of modules for doing specialized tasks. There are routines for manipulating images, for calculating fast Fourier transforms, for encrypting data, etc. So these general-purpose languages can be very useful for doing mathematics.

These languages are especially convenient for programmers who have to deal with data existing as text. All have strong built-in functionality for dealing with strings of text. For example, they can easily be programmed to detect lines of a file containing certain words or to substitute one word for another in data.

Web pages are one familiar example of text as data. Web pages are just carefully structured text files, and consequently Perl is the most common language for dynamic Web programming (although all three are popular with Web developers).

As pure programming languages, Python and Ruby are perhaps a bit "fuller" than Perl, sometimes in aspects that might be important to a mathematician. In particular, three-dimensional arrays are not conveniently represented in Perl, and for some kinds of work this could be a stumbling block.

It is difficult to recommend one language over another among these three. In terms of popularity, Perl is more popular than Python, which is more popular than Ruby. In terms of elegance, Ruby is arguably more elegant than Python, which is more elegant than Perl. Of course, elegance is a subjective concept and popularity does not get your individual project done!

Java, C, and Fortran

Java is a high-level language known for its cross-platform nature (the same program can run on Mac OS X, Microsoft Windows, and GNU/Linux). C and Fortran are mid-level languages known for their speed. Historically, Fortran is one of the oldest computer programming languages and today it is less popular than C, but it is very powerful and still finds use for new mathematical work.

Generally, high- and mid-level languages become a consideration when math-specific or very high-level languages have some shortcoming that affects your project. For example, your program may need copious amounts of memory to perform its work. In a very high-level language, the language handles much of the duty of memory management. In a mid-level language, like C, this becomes the task of the programmer. It is less convenient, but it also translates into more control and possibly more success.

Mid-level languages, particularly C and Fortran, are also known for their speed. It may be that your project requires extensive processing, and a very high-level language is too slow. C and Fortran are compiled directly into binary code that the computer natively understands, and this usually makes for very fast programs.

Note. Java is also compiled, but not into native code. Rather, Java is converted into a kind of "ideal" binary code called bytecode that will run the same on any computer no matter what kind of processor it uses. This is convenient for portability, since the same program can run on a Mac or PC, but it is not as fast as good C or Fortran code can be.

Most of the programs you are familiar with were written in Java, C, or the "C-like" language, C++. If you are writing code that will become a polished stand-alone product, then you will probably prefer one of these languages as well.

Other languages

The lowest-level language is assembler language. Programs written in assembler are crafted using the instructions of a particular type of processor, and essentially all aspects of the program are under the control of the author. This kind of programming is not for beginners and usually not for large programs. But for certain kinds of specialized programming, no other language can compare for speed or flexibility.

There are numerous languages that could be suitable for projects. Lisp is one of the oldest computer languages, though not quite as old as Fortran, and is still used today. Pascal was popular in the 1980s and is still a fine language. BASIC is not as popular today as it was in the 1970s, but it remains a capable language.

Perhaps JavaScript deserves a special note, particularly for programs you intend to share. JavaScript is a computer language that is part of Web browsers, which makes JavaScript programs accessible for sharing in a way that no other stand-alone program can be. If your program can be expressed in JavaScript, then it can be shared with anyone who has a Web browser. Users do not even have to install anything; they simply visit a Web page containing your JavaScript program and run it.

Consider communities

The technical advantages or disadvantages of a language are not necessarily the most important considerations when starting a project. There are many communities that you may want to connect with, and this may influence or dictate your choice of language. Statisticians may want to consider using R not only because of its capabilities, but because it is a standard in the field. Other statisticians are likely to be familiar with R, and there is a body of work to draw from and contribute to.

Similarly, although image processing often requires the kind of speed that might lead you to consider a fast compiled language such as C, Java might in fact be a more natural choice. There is a large collaborative project called ImageJ written in Java. Basing your own work on ImageJ will make it easier to get started, and good work that you produce can be contributed back for other researchers to use.

Research related to defense or national security might naturally be written in Ada, not only because of its technical merits, but because it is a historically important language in the defense industry.

Finally, the free software community should not be overlooked. There are thousands of programs available as free software that come ready to use and with source code to study. Many are associated with the GNU/Linux operating system, but a number of others are also available for Microsoft Windows or Mac OS X. A student may well find software to use or adapt for their own projects that comes from a free software project.

Because so much free software is written for GNU/Linux (and Unix) operating systems, C is a great language for students to learn if they have a special interest in free software. Linux was written in C, as were the standard tools of Linux. The 'C library' essentially connects all software that runs in a Linux environment (no matter the language), so understanding C can make programming easier even if your preferred language is something else.

In summary, you can usually benefit by trying to learn from the work of others, and you will want a way to contribute or share the work that you do. Look for an appropriate community to become part of and consider using the tools most popular with those people.

18.3   How to learn more

Documentation for each of the major mathematical software programs, including use as a programming language, is available through its respective Web site:

- Mathematica: `www.wolfram.com`

- Maple: `maplesoft.com`

- Maxima: `maxima.sourceforge.net`

- MATLAB: `www.mathworks.com`

- Octave: `www.gnu.org/software/octave`

- R: `www.r-project.org`

Generally, each site links also to printed documentation available for purchase.

Since R is an implementation of the S programming language, books on S may be useful in addition to the documentation on the R project Web site.

For Perl programming, the O'Reilly Media book *Learning Perl* [42] by Tom Phoenix, Randal L. Schwartz, and brian d foy is without peer. It will teach you not only how to program in Perl, but also how to think in Perl. The similarly titled *Learning Python* by Mark Lutz [35] (also published by O'Reilly Media) is a good resource for Python programming.

Large Internet communities have grown around many free computer programming languages, and these can be valuable resources when you are looking for code to use or examples to study. Among the biggest and most vibrant projects are:

- Comprehensive Perl Archive Network: `www.cpan.org`

- Python Package Index: `pypi.python.org`

- RubyForge: `rubyforge.org`

- Comprehensive R Archive Network: `cran.r-project.org`

- Comprehensive TEX Archive Network: `ctan.org`

For any popular programming language, it is smart to visit your local libraries. Most libraries will have books on C or C++, Java, JavaScript, and probably other languages. You should be able to find books that can introduce you to your preferred language or help you build up your skills.

Exercises

1. Check out a book from the library on a programming language you have never used, and write enough programs to start getting comfortable with the syntax.

2. Write an implementation of the Euclidean algorithm for finding the greatest common divisor of two numbers (this is discussed briefly on p. 122). Test it with positive numbers, negative numbers, and zero to see if it always gives a correct result.

3. Write an implementation of the binary GCD algorithm, an alternative to the Euclidean algorithm. The binary GCD algorithm uses division by 2 (which is a fast operation on binary computers) to find $\gcd(a, b)$ when a and b are nonzero. The steps are:

   (i) If a and b are both even, divide both by 2 until one of them becomes odd. Remember the number of 2's that are removed (because they are part of the GCD).

   (ii) Divide out any remaining factors of 2 from a or b and discard them. Now both a and b are odd.

   (iii) Swap a and b, if necessary, to make $a \geq b$.

   (iv) Replace a with $a - b$.

   (v) If $a = 0$, then b is the odd part of the GCD. Combine with the even part from step (i) and return the result.

   (vi) Otherwise, a is even (because we subtracted two odd numbers). Throw away all the factors of 2 and go back to step (iii).

4. Write a program that does the calculation to estimate an area under a curve using the Midpoint rule (also discussed in Exercise 6 on p. 125). Write your program so that it is easy to change the function or the number of rectangles.

5. Write a program that does the calculation to estimate an area under a curve using Simpson's rule. Write your program so that it is easy to change the function or the number of intervals.

6. Write a program that calculates the slope and y-intercept of the line of best fit for a set of points. Write your program so it is easy to change the points (or have it ask the user to enter the points).

Chapter 19

Getting Started with Free and Open Source Software

In this chapter, we explore the concept of free software and discuss several popular applications for mathematics.

19.1 What is free and open source software?

Free software is an idea proposed by Richard Stallman that is based on the premise that our software world is better when users of software have more freedom. It's a happy coincidence that free software is usually available at no cost, but that is not the primary point. The primary idea is freedom.

In order for a piece of software to be free software, its license should grant users the following freedoms [54, p. 41].

- The freedom to run the program, for any purpose (freedom 0).

- The freedom to study how the program works, and change it to make it do what you wish (freedom 1). Access to the source code is a precondition for this.

- The freedom to redistribute copies so you can help your neighbor (freedom 2).

- The freedom to distribute copies of your modified versions to others (freedom 3). By doing this you can give the whole community a chance to benefit from your changes. Access to the source code is a precondition for this.

The Free Software Foundation (`fsf.org`) is the steward of the idea of free software. It develops and helps defend the most popular free software license, the General Public License.

Many of the freedoms in the definition of free software depend on users having access to source code. Users cannot realistically study or modify software unless they have source, for example.

"Open source" is a software philosophy that is similar to free software, but the open source philosophy focuses more on development process than freedom. According to the open source philosophy, when the source code for a program is open and can be studied, software develops more quickly and is of higher quality. The Open Source Initiative (`opensource.org`) is the primary promoter of the open source philosophy.

Free and open source software fit well philosophically into the academic world, since we expect new knowledge to build on old knowledge, and we expect new discoveries to be shared and improved upon.

19.2   Why use free and open source software?

Instinct would tell us that software that can be obtained for free cannot be of very high quality. As the cliché goes, "There's no such thing as a free lunch." But this turns out to be incorrect. Many good pieces of software that mathematics students will be interested in are free software. Many of the software programs discussed in this book are available under free or open source software licenses, including:

- Firefox

- TeX and LaTeX

- PSTricks and Beamer

- Maxima, Octave, and R

- Ghostscript (an implementation of PostScript)

- Most programming languages, including C, Perl, Python, Ruby, Java, and Fortran

There are also other free and open source programs devoted to various mathematical topics. Examples include:

- Axiom, Yacas, general computer algebra systems

- GAP, for Computational Discrete Algebra

- gnuplot, for plotting and visualization

- Macaulay2, for Algebraic Geometry and Commutative Algebra

- PARI/GP, for number theoretic computations

- Sage, a "broad" mathematical software system

These are just some of the larger mathematical software projects. There are easily hundreds, probably thousands, of other small and large projects that will be useful or interesting to mathematics students.

Most free software programs will run on a variety of systems, including Microsoft Windows, Mac OS X, and Unix (including free software Unix systems like BSD as well as proprietary systems like IBM's AIX). But there is one operating system that deserves special mention. No discussion of free and open source software would be complete without talking about one of the most successful free software projects ever, GNU/Linux.

19.3 What is Linux?

Linux[1] is an operating system. Operating systems are computer programs, but they are a bit different from the application programs that we usually think of when we use the word "programs." Most people are familiar with only two operating systems, Microsoft Windows and Mac OS X, but they usually know of many many application programs.

The primary purpose of an operating system, from the viewpoint of a user, is to determine the way the computer functions in a broad generic way. The operating system manages the individual user accounts on the computer and provides security, perhaps requiring you to enter a password before you can use the computer. It also provides a way for you to organize your files, allowing you to copy files from place to place, and letting you delete files.

Generally speaking, it is the operating system you work with when you are at the computer but not using a specific application (like a word processor). When you "drag" files from a CD-ROM to your hard drive, you are using the operating system. In contrast, when you enter data into a spreadsheet, the operating system is running, but your focus is on an individual application: the spreadsheet program itself.

An operating system also does something important from the point of view of a computer programmer. It manages the hardware of the computer. Programmers don't want to know the differences between a hard drive made by Western Digital and a hard drive made by Seagate. Programmers don't need to know the differences between a network card made by Linksys and a network card made by Cisco. So the operating system manages the hard drives, the network cards, and the other hardware in the computer. This happens via device drivers. When you install a driver, you are adding something to the operating system that makes some piece of hardware work.

Most of the work of an operating system happens somewhat "under the hood," making the hardware work. This leaves us to do the things we want to do: applications. To some extent, we don't care about operating systems any more than we care about what kind of engine is in our car. We care that our car works and runs, and the same is true for operating systems. We care that they run our word processors, Web browsers, or spreadsheets.

Yet there is something different between a car with a four-cylinder engine and a car with a V8, even if both will effectively drive you to work. In the same way, we often do care about operating systems, because they define the personality of our computers. Enthusiasts of the Apple Macintosh line are devoted to it in part because of the way Mac OS X manages

---

[1]Technically, Linux is the name of the operating system kernel written by Linus Torvalds and a global team of contributers. The Free Software Foundation espouses the full name "GNU/Linux" to refer to the kernel and tools together that make a fully working operating system. Common usage is to abbreviate the name to "Linux," so throughout this book, you should interpret "Linux" to mean "GNU/Linux."

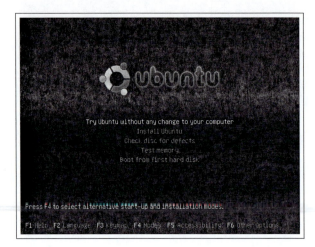

FIGURE 19.1: Booting an Ubuntu installer CD.

everything. It just feels different to use a Mac than to use a PC running Windows, even when you are doing the same tasks on each computer.

Just as some people will enjoy the "fit" of Mac OS X, others will connect with a third popular operating system, Linux. Linux was originally created by programmers for themselves, so often people of a technical persuasion find themselves at home with Linux. However, as Linux has become more popular, it has become easier to use and now many non-programmers are attracted to Linux.

19.4 How to install Linux

Linux can be acquired for free, and copies can be downloaded via the Internet from sites all around the world. Numerous organizations and companies have worked to put together Linux operating systems and related applications, packaged as collections called "distributions." Most Linux distributions cost nothing, although some are available for purchase and come with commercial support. Popular distributions include Ubuntu, Redhat, Debian, Suse, and Slackware, and there are numerous others. Distributions are the key to getting Linux installed on your computer.

For beginners, the Ubuntu Linux distribution (see Figure 19.1) is probably a good choice. Ubuntu was created by a company called Canonical, which has invested heavily to put together a Linux distribution that works well for first-time Linux users. Installation CDs can be downloaded from ubuntu.com, and they can be ordered through the mail as well.

The standard Ubuntu CD (the "desktop" version) has a feature that lets you try Linux without risk. It is a "Live" CD, meaning that you can reboot your computer with the Ubuntu CD in the CD drive, and Linux will start straight from the CD (see Figure 19.2). Ubuntu will detect your hardware and run without permanently installing anything and without harming your existing data. If Linux runs well from the Live CD, you can have confidence that it will run well when you install it permanently on your computer.

Installing Linux on your computer usually involves three steps:

1. Creating empty space on the hard drive for Linux.

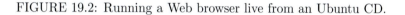

FIGURE 19.2: Running a Web browser live from an Ubuntu CD.

2. Copying the Linux operating system and applications to the hard drive.

3. Configuring the computer to boot Linux each time the computer starts.

To make room for Linux, it is simplest to completely wipe the drive and install Linux in the empty space, but this will erase all of your existing files and programs. Often, people don't want to give up their current system, even when they are eager to switch to Linux. That is fine. As long as you have not completely filled your hard drive with data, the Linux installer can generally shrink what you have—nondestructively—and install Linux next to it. "Shrinking" is a pretty intensive operation, though, so you would be wise to back up your data before trying it (although it is almost always safe).

While it's possible to have Linux on the same hard drive as another operating system, like Microsoft Windows or Mac OS X, it isn't possible for Linux and the other OS to simultaneously control the computer.[2] Remember: operating systems manage the hardware at the lowest levels. They cannot share this job; one or the other has to do it. The most common way to handle this is to have the computer ask at boot time if you would like to run Linux or your other operating system. Once you have chosen, the selected OS will run the computer until the next reboot.

19.5 Where to get Linux applications

One of the common mistakes that new Linux users make is looking in the "wrong" place for applications, for the actual programs they intend to use from day to day (like Web browsers and word processors). It is often *not* necessary to go looking for programs on the Internet to download. Hundreds of applications have already been prepared to work with Linux and they are part of most distributions. The user simply needs to know how to install the applications she wants.

---

[2]If you really need to run two operating systems at once, an emulator like VirtualBox, QEMU, DOSEMU, or VMware may be for you.

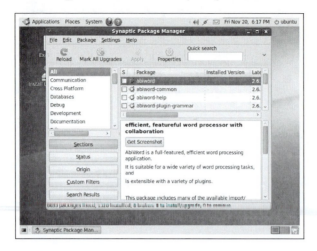

FIGURE 19.3: Installing packages with Synaptic.

A user of Microsoft Windows or Mac OS X could be forgiven for not knowing that Linux distributions already contain a plethora of applications, since it works somewhat differently with proprietary operating systems. Numerous Web sites are devoted to "freeware" or "shareware" programs for Microsoft Windows. But because so much software comes ready to install with a Linux distribution, Linux users only rarely install software they have downloaded from somewhere else.

For example, the Firefox Web browser is the dominant browser among Linux users. While Firefox can be downloaded from its primary Web site, every common Linux distribution already has it built in. You probably have Firefox already installed without even asking for it. If you do not, you can simply pick it from a menu and it will be downloaded automatically.

On an Ubuntu system, new programs can be installed literally with a click of the mouse. At the top of the screen there is a **System** menu that has a tab called **Administration**. Under **Administration** is a program called **Synaptic Package Manager**, often referred to simply as Synaptic (see Figure 19.3). Synaptic maintains a database of software that has been customized for your Ubuntu system. Installing a new program is as simple as checking a few boxes and clicking **Apply**; Synaptic handles the details. Synaptic is pretty intelligent. For example, if one program depends on another to work correctly, Synaptic knows that and will help you install both.

19.6   How is Linux familiar?

If we remember that a big part of operating systems is to do work behind the scenes, then it is not surprising that Linux systems work and feel a lot like PCs and Macs. The same applications tend to work about the same no matter what operating system you are using. If you run Firefox on Linux, it's pretty much the same as Firefox on Windows or Firefox on Mac OS X. If you run a word processor on Linux, it will feel a lot like a word processor on any other computer.

All of the common types of application are available in some form for Linux. OpenOffice is available for word processing, spreadsheets, and presentations. Firefox is available for browsing the Web. There are computer languages like Java, Perl, and C if you are interested in being a programmer. There are image programs like the GIMP for editing pictures. All of these programs run essentially the same no matter what operating system sits beneath them.

Many applications of interest to mathematicians are available for Linux. LaTeX, Beamer, Maxima, Octave, and R are free and almost certainly packaged to work with your favorite distribution. In addition, Mathematica, Maple, and MATLAB are available for purchase, and they work the same under Linux as on other systems. Linux can be a capable system, running your favorite tools.

19.7 How is Linux different?

Differences in common software

Probably the first thing new users will miss on a Linux system is Microsoft Office. There is no Microsoft Word, no Excel, no PowerPoint, etc. Microsoft does not produce a version of Office for Linux; you cannot purchase it or download it. For some people this is a deal breaker. For others, a substitute program like OpenOffice may be a perfectly acceptable alternative.

Linux is also different from Windows and Mac OS X in an important historical way, one that has some of its roots in a system called Unix. Unix was written by programmers for their own use, and aspects of the Unix system reflect that, particularly reliance on the "command line" interface. Similarly, Linux users are usually comfortable typing commands to control their computers, as were the original Unix devotees. In fact, Linux users often embrace the command line, using commands to do tasks that Windows users or Mac users would choose to do with the mouse.

To some extent, Linux continues to be designed around this propensity. It is not only possible, but it is common practice to open a "terminal" window and start issuing commands (see Figure 19.4). On an Ubuntu system, terminals can be started from the Applications/Accessories menu (and there may even be more than one choice on the menu—terminals are that important).

Terminal windows will be familiar and at the same time foreign to Windows users. They work essentially like the common "DOS box" that Windows users know as cmd.exe. However, the Linux terminal is more capable, and the manner in which it is used is also different. While PC users will typically only type a command or two in a DOS box as needed, it would not be uncommon for a Linux user to spend most of a session working in one.

Mac users may adapt more quickly when it comes to using the terminal because Macs use some of the same software that Linux systems use. The most popular command line interface on Linux systems is provided by a program called bash, and this is also the program that Macs use in their terminal windows. So the advanced Mac user who has already used terminal windows should find the Linux command line familiar and predictable.

Differences in storing files

When it comes to file storage, Linux systems are different from both Macs and Windows computers, although perhaps the differences encountered by former Windows users are

FIGURE 19.4: A terminal window waiting for commands.

greater. One simple difference is almost cosmetic, however. Windows computers use the backslash (\) to indicate folders and sub-folders. Like the Mac, and like World Wide Web addresses, Linux uses forward slash (/). It is also a bit more common to refer to folders and sub-folders by the terms "directories" and "subdirectories" when working on Linux systems, but the concept is the same.

A more notable difference is that file names on Linux systems are case sensitive. The files Don.txt, DON.txt, and DON.TXT are all different files and could all exist simultaneously in the same directory on a Linux system. Neither Windows nor Mac OS X (with the default configuration) would be happy with a scenario like that.

Another difference that Windows users are bound to find odd is that there are no drive letters on Linux systems; in particular there is no drive C:. Indeed, there is essentially no analogy to drive letters at all. USB storage devices (commonly called jump drives) do not have a drive letter, and neither do CD-ROMs or DVDs.

Instead, all files and folders are grafted together into a single tree descending from the top or "root" directory /. For example, when you insert a CD-ROM into the CD drive, all of its files and folders might be found under the name /media/cdrom0. Similarly, a jump drive will have a folder name that represents it and contains its files.

In practice, the precise name a CD-ROM gets when inserted into the computer often does not make much difference. Most modern distributions will place an icon on the desktop when you insert a new CD-ROM or jump drive, and each device can be opened by double-clicking on it.

Personal files, like the files a Windows user might save in "My Documents," are traditionally stored in directories under the /home directory, typically named by the login name of each account. For example, if your login name is goodstudent, you can expect your personal files to be in /home/goodstudent. This means, for instance, that the files on your desktop probably reside in /home/goodstudent/Desktop.

Differences of philosophy

Linux is a compelling platform for users of free and open source software (including mathematical software).

Nearly all of the software that comes with a Linux distribution is free software. An important reason that Firefox can come built-in to Ubuntu is because it is free (in the freedom sense). The Ubuntu programmers have the freedom to take Firefox, modify it for Ubuntu, and then give you copies. The same is true for hundreds of other programs tracked in the Synaptic database. If you are interested in using some free software program, it has probably already been adapted and tested by Ubuntu developers and volunteers, and incorporated into the distribution. All you need to do is select it and use it.

As a math student or math professional, you could download and install most free math software for a system running a proprietary system like Mac OS X or Microsoft Windows. But it's potentially more streamlined and seamless to run that kind of software on a Linux system. On Windows, for example, programming languages are an afterthought, whereas most Linux systems have Perl installed by default, and many have C and Python. Free software not installed by default is generally straightforward to add to the system (because it has usually been packaged and integrated into your Linux distribution).

All of the software listed in Section 19.2 is available for Linux. Only Sage has not been completely incorporated into the Ubuntu distribution, and even it has Ubuntu-compatible files available for downloading on its Web site. It's likely that future versions of Ubuntu will simply contain Sage.

One beautiful (and perhaps unexpected) consequence of the free software philosophy is that Linux systems are essentially virus free. In practice, Linux users rarely if ever install unvetted "potentially suspicious" software. There is little incentive to download an application (that might contain a virus) from a dubious Internet Web site when hundreds of safe free programs are waiting to be installed with the click of a mouse.

The beauty of differences

It is the differences that make people prefer one particular program over others, and it is also differences that draw people to Linux over other capable systems like Microsoft Windows and Mac OS X. Maybe you find yourself drawn to the power and elegance of the command line interface. Perhaps you are intrigued or inspired by the concept of free software. Or maybe you just want to try something a bit different. Whatever the reason, Linux might be for you.

19.8 How to learn more

To learn more about the philosophy of free software, visit the main site of the Free Software Foundation, (`fsf.org`). The GNU project (`gnu.org`) may also be of interest to you. There you will learn that although Linux gets the fanfare for being the most successful free software operating system, many of the tools and programs that we think of as part of Linux actually come from the GNU project.

The main site for Ubuntu Linux is `ubuntu.com`. Also worth checking out are the Ubuntu Forums (which contain discussions about Ubuntu and a place to ask questions) at `ubuntuforums.org`.

The definitive Web sites for the software listed in Section 19.2 are:

- Axiom: `axiom-developer.org`

- Firefox: `getfirefox.com`

- GAP: `gap-system.org`

- gnuplot: `gnuplot.info`

- Macaulay2: `www.math.uiuc.edu/Macaulay2/`

- PARI/GP: `pari.math.u-bordeaux.fr`

- Sage: `sagemath.org`

- Yacas: `yacas.sourceforge.net`

SourceForge (`sourceforge.net`) is a huge site devoted to the support of open source software. Many free and open source software projects have their project homes stored there, and they can be located through the search feature. Freshmeat (`freshmeat.net`) is a kind of searchable clearinghouse for software, where authors can post announcements. Many projects are registered with the Freshmeat Web site, including numerous free and open source programs. Both SourceForge and Freshmeat are good starting points when searching for free and open source software, mathematical or otherwise.

For further reading, consult the writings of Richard M. Stallman, which have defined the philosophy of free software and guided the free software movement for two decades. They are conveniently compiled in [54]. Other books that have helped inspire and shape the free software culture are [44] and [34].

Exercises

1. Request a free CD copy of Ubuntu Desktop Edition from Canonical Ltd., or download an image from the Ubuntu Web site and burn a CD (or DVD) copy yourself. Boot your computer using the Ubuntu disc as a "live disc" to try Ubuntu without deleting your existing software. Hint: There are "getting started" directions on the Web site that include help on how to burn a disc.

2. Visit `sagemath.org`.

   (a) Find out what license Sage uses.

   (b) Find out what (free and open source) programs make up the components of Sage.

   (c) Browse several of the available Sage notebooks.

   (d) Create an account, and try Sage via the Web interface. Hint: There is a tutorial on the Web site that will help you get going.

3. Search `freshmeat.net` and `sourceforge.org` and describe three promising free or open source programs for each topic. Make sure you note the license that each piece of software uses.

   - Linear Algebra
   - Graph Theory
   - Differential Equations
   - Statistical Analysis

4. Read the General Public License, as published by the Free Software Foundation. Read either the current version, or the significantly shorter and simpler version 2.0.

   • `www.gnu.org/licenses/gpl.html`, current version

   • `www.gnu.org/licenses/old-licenses/gpl-2.0.html`, version 2.0

5. Visit Project Gutenberg at `gutenberg.org`.

   (a) Describe the purpose of the project.

   (b) Search the project for books on mathematics or related subjects. Hint: The Advanced Search feature allows searching via Library of Congress classifications.

   (c) Describe how to volunteer for the project through "distributed proofreading."

6. Search the Internet for free or open source math books. Hint: You will need to check the licenses, since not all "free" books have free licenses.

Chapter 20

Putting it All Together

When we do mathematical work, we have many tools to choose from. Much of mathematical work is of three types:

- Text

- Pictures

- Calculations

In the area of text, the best choice for most writing is LaTeX (see Chapter 9). LaTeX is the professional standard all over the world. When people exchange documents, work on collaborative projects, submit articles to journals, etc., the work is invariably composed in LaTeX. However, if you want your text to appear on the World Wide Web, then you would compose it in HTML (Chapter 15). On your Web page, you may want to link to LaTeX documents, or LaTeX documents converted to PDF (Portable Document Format).

When making pictures, there are many choices, depending on the situation. If the picture is a simple addition to your LaTeX document, then LaTeX's `picture` environment may suffice. If the picture is more complicated, especially if it involves plotting points, then you may want to use PSTricks (Chapter 10), which is invoked as a package within your

242   20 Putting it All Together

LaTeX document. If your picture is complex, especially involving programming constructs, then you may want to use PostScript (Chapter 17). A final option is to create the image using a computer algebra system (Chapters 12 and 13).

When it comes to doing calculations, we have seen that the options range from using a computer algebra system to writing a program in the computer programming language of your choice. As we saw in Chapter 18, there are many options as to what programming language one could choose.

Exercises

In the following exercises, you are invited to choose which software tools to use and how to use them.

1. Typeset a homework assignment. Include any necessary pictures.

2. Write a short article on an aspect of mathematics history. Make your work available to others on the World Wide Web. Be sure to cite your references and provide links to useful resources.

3. Give a presentation on the mathematical figure known as Torricelli's Trumpet.

4. Give a presentation on the seven frieze groups.

5. Give a presentation on Jacobi's identity. What are some familiar contexts in which the identity holds? What is the importance of the identity in mathematical physics?

6. Look up the dodecahedral graph and draw a picture of it.

7. Make the following diagram.

(The diagram represents 84 cards without a set in the SET game generalized to 6 dimensions.)

8. Recall the squaring map described on p. 70. Make a picture of the squaring map defined on the integers modulo 25.

9. The graph shown in Figure 14 is a two-colored complete graph on 17 vertices. The color of an edge from i to j is defined based on whether $i - j$ is a nonzero square or a nonsquare modulo 17. The nonzero squares modulo 17 are $\{1, 4, 9, 16, 8, 2, 15, 13\}$.

   Draw a three-coloring of the edges of the complete graph on 16 vertices, with the property that there is no triangle all of whose edges are the same color.

   Any coloring of the edges of a complete graph on 17 vertices, using three colors, must contain a triangle with all three edges the same color. Can you prove this result, which comes from the field of Ramsey Theory?

10. Make a program to produce Pascal's triangle and output the number of odd entries in each row of the triangle. What is the pattern?

11. Make a program to count the number of ways to make change for a dollar, using any number of pennies, nickels, dimes, quarters, and half-dollars.

12. Let ABC be an equilateral triangle of side length s. Let P be a point in the interior of ABC. Suppose that the distances from P to A, B, C are a, b, c, respectively. Find a quadruple of values $\{s, a, b, c\}$ that are all integers.

13. Recall the power mean as defined in Example 9.5. Choose two distinct positive real numbers x_1 and x_2, and let $\mathbf{x} = (x_1, x_2)$. Graph the rth power mean $M_r(\mathbf{x})$, for $-10 \leq r \leq 10$. Can you make a conjecture about the shape of the power mean graph?

14. Recall the Vandermonde matrix as defined in Example 9.5. Let V_n be the Vandermonde matrix of order n. Verify that

$$\det V_6 = \prod_{1 \leq i < j \leq 6} (x_j - x_i).$$

15. (The squaring map revisited.) Define a directed graph as follows. The vertices are the integers 0, 1, 2, ..., $10^6 - 1$ (one million vertices). Vertex A is directed to Vertex B if and only if $A^2 \bmod 10^6 = B$. We say that Vertex A and Vertex B are in the same component of the directed graph if there is a directed path from A to B or from B to A. How many components are there? What is the length of a longest directed cycle?

Bibliography

[1] Adobe Systems Incorporated. *PostScript Language: Program Design.* Addison-Wesley, Reading, MA, 1985.

[2] Adobe Systems Incorporated. *PostScript Language: Reference Manual.* Addison-Wesley, Reading, MA, 1985.

[3] Adobe Systems Incorporated. *PostScript Language: Tutorial and Cookbook.* Addison-Wesley, Reading, MA, 1985.

[4] M. Aigner and G. M. Ziegler. *Proofs from THE BOOK.* Springer-Verlag, New York, third edition, 2004.

[5] D. J. Albers and G. L. Alexanderson, editors. *Mathematical People: Profiles and Interviews.* Burkhäuser, Boston, first edition, 1985.

[6] D. J. Albers, G. L. Alexanderson, and C. Reid, editors. *More Mathematical People: Contemporary Conversations.* Academic Press, Providence, RI, first edition, 1990.

[7] B. Casselman. *Mathematical Illustrations: A Manual of Geometry and PostScript.* Cambridge University Press, New York, 2005.

[8] E. Castro. *HTML for the World Wide Web: Visual Quickstart Guide.* Peachpit Press, Berkeley, CA, fourth edition, 2000.

[9] E. Castro. *XML for the World Wide Web: Visual Quickstart Guide.* Peachpit Press, Berkeley, CA, first edition, 2001.

[10] D. Cederholm. *Web Standards Solutions: The Markup and Style Handbook.* Apress, New York, 2004.

[11] K. R. Coombes, B. R. Hunt, R. L. Lipsman, J. E. Osborn, and G. J. Stuck. *The Mathematica Primer.* Cambridge University Press, New York, 1998.

[12] A. Diller. *LaTeX: Line by Line: Tips and Techniques for Document Processing.* John Wiley & Sons, New York, second edition, 1999.

[13] C. Ferguson. *Helaman Ferguson: Mathematics in Stone and Bronze.* Meridian Creative Group, Erie, PA, 1994.

[14] B. A. Fusaro and P. C. Kenschaft, editors. *Environmental Mathematics in the Classroom.* Mathematical Association of America, Washington, DC, 2003.

[15] S. Glaz and J. Growney, editors. *Strange Attractors: Poems of Love and Mathematics.* A. K. Peters, New York, 2009.

[16] J. Glynn and T. Gray. *The Beginner's Guide to Mathematica.* Cambridge University Press, New York, fourth edition, 2000.

[17] M. Goossens, F. Mittelbach, and A. Samarin. *The LATEX Companion.* Addison-Wesley, Reading, MA, second edition, 2000.

[18] M. Goossens and S. Rahtz. *The LATEX Web Companion: Integrating TEX, HTML, and XML.* Addison-Wesley Series on Tools and Techniques for Computer Typesetting. Addison-Wesley, Reading, MA, 1999.

[19] M. Goossens, S. Rahtz, and F. Mittelbach. *The LATEX Graphics Companion: Illustrating Documents with TEX and PostScript.* Addison-Wesley Series on Tools and Techniques for Computer Typesetting. Addison-Wesley, New York, 1997.

[20] G. Grätzer. *First Steps in LATEX.* Birkhäuser, Boston, 1999.

[21] D. F. Griffiths and D. J. Higham. *Learning LATEX.* SIAM, New York, 1997.

[22] R. Haberman. *Mathematical Models.* Society for Industrial and Applied Mathematics, Philadelphia, 1998.

[23] C. R. Hadlock, editor. *Mathematics in Service to the Community.* Mathematical Association of America, Washington, DC, 2005.

[24] Margie Hale. *Essentials of Mathematics: Introduction to Theory, Proof, and the Professional Culture.* Mathematical Association of America, Washington, DC, 2003.

[25] N. J. Higham. *Handbook of Writing for the Mathematical Sciences.* Society for Industrial and Applied Mathematics (SIAM), Philadelphia, 1998.

[26] S. Hollis. *Multivariable CalcLabs with Mathematica for Stewart's Multivariable Calculus.* Brooks/Cole, New York, fourth edition, 1999.

[27] S. Hollis. *Single Variable CalcLabs with Mathematica for Stewart's Calculus, Single Variable.* Brooks/Cole, New York, fourth edition, 1999.

[28] S. Kalajdzievski. *Mathematics and Art: An Introduction to Visual Mathematics.* CRC Press, Boca Raton, FL, 2008.

[29] D. E. Knuth. *The TEX Book.* Addison-Wesley, Reading, second edition, 1986.

[30] D. E. Knuth, T. Larrabee, and P. M. Roberts. *Mathematical Writing.* Mathematical Association of America, Washington, DC, revised edition, 1989.

[31] S. G. Krantz. *A Primer of Mathematical Writing: Being a Disquisition on Having Your Ideas Recorded, Typeset, Published, Read, and Appreciated.* American Mathematical Society, Providence, RI, 1997.

[32] S. G. Krantz. *How to Teach Mathematics.* American Mathematical Society, Providence, RI, second edition, 1999.

[33] L. Lamport. *LATEX: A Document Preparation System, User's Guide and Reference Manual.* Addison-Wesley, New York, second edition, 1994.

[34] L. Lessig. *How Big Media Uses Technology and the Law to Lock Down Culture and Control Creativity.* Penguin Press, New York, 2004.

[35] M. Lutz. *Learning Python: Powerful Object-Oriented Programming.* O'Reilly, Cambridge, MA, fourth edition, 2009.

[36] R. E. Maeder. *Computer Science with Mathematica: Theory and Practice for Science, Mathematics, and Engineering.* Cambridge University Press, New York, first edition, 2001.

[37] E. Martin and J. Novak, editors. *Mathematica 4 Standard Add-on Packages.* Cambridge University Press, New York, first edition, 1999.

[38] H. McGilton and M. Campione. *PostScript by Example.* Addison-Wesley, Reading, MA, 1992.

[39] M. Mesterton-Gibbons. *A Concrete Approach to Mathematical Modeling.* John Wiley & Sons, New York, 1995.

[40] A. Navarro and T. Stauffer. *HTML by Example.* QUE, Indianapolis, IN, first edition, 1999.

[41] I. Peterson. *Fragments of Infinity, A Kaleidoscope of Math and Art.* John Wiley & Sons, Inc., New York, 2001.

[42] T. Phoenix, R. L. Schwartz, and b. d. foy. *Learning Perl.* O'Reilly, Cambridge, MA, fifth edition, 2008.

[43] G. Pólya. *How to Solve It: A New Aspect of Mathematical Method.* Doubleday, New York, 1957.

[44] Eric S. Raymond. *The Cathedral and the Bazaar: Musings on Linux and Open Source by an Accidental Revolutionary.* O'Reilly, Cambridge, MA, 1999.

[45] G. C. Reid. *Thinking in PostScript.* Addison-Wesley, Reading, MA, 1990.

[46] N. Rodgers. *Learning to Reason: An Introduction to Logic, Sets, and Relations.* Wiley, New York, 2000.

[47] K. Rosen. *Discrete Mathematics.* Pearson/Prentice Hall, New York, sixth edition, 2011.

[48] H. Ruskeepää. *Mathematica Navigator: Graphics and Methods of Applied Mathematics.* Academic Press, New York, first edition, 1998.

[49] D. G. Sarvate and J. Seberry. Suggestions for presentations of a twenty-minute talk. *Bulletin of the Institute of Combinatorics and its Applications*, 4:90–91, 1992.

[50] P. Schiavone. *How to Study Mathematics: Effective Study Strategies for College and University Students.* Prentice Hall, Upper Saddle River, NJ, 1998.

[51] D. Smith, M. Eggen, and R. St. Andre. *A Transition to Advanced Mathematics.* Brooks/Cole, Chicago, seventh edition, 2009.

[52] R. Smith. *PostScript: A Visual Approach.* Peachpit Press, Berkeley, 1990.

[53] W. Snow. *TEX for the Beginner.* Addison-Wesley, Reading, 1992.

[54] R. M. Stallman. *Free Software Free Society: Selected Essays of Richard M. Stallman.* GNU Press, Boston, 2002.

[55] E. Swanson, A. O'Sean, and A. Schleyer. *Mathematics into Type.* American Mathematical Society, Washington, DC, updated edition, 1999.

[56] B. F. Torrence and E. A. Torrence. *The Student's Introduction to Mathematica: A Handbook for Precalculus, Calculus, and Linear Algebra.* Cambridge University Press, New York, 1999.

[57] D. J. Velleman. *How to Prove It: A Structured Approach.* Cambridge University Press, New York, second edition, 2006.

[58] S. Wagon. *Mathematica in Action.* Springer-Verlag, New York, second edition, 1999.

[59] S. Wolfram. *The Mathematica Book.* Cambridge University Press, New York, fifth edition, 2003.

Index